OUT OF THE ENERGY LABYRINTH

OUT OF THE ENERGY LABYRINTH

*Uniting Energy and the Environment
to Avert Catastrophe*

As power prices gyrate, energy uncertainties mount,
global warming dangers increase and the oil-rich
Middle East sinks into deeper chaos – a guide to the
maze of myths, illusions, fears, hopes, personalities,
events and opportunities which shape the twinned
debates on energy security and climate security.

By
David Howell
and
Carole Nakhle

I.B. TAURIS
LONDON · NEW YORK

Published in 2007 by I.B.Tauris & Co Ltd
6 Salem Road, London W2 4BU
175 Fifth Avenue, New York NY 10010
www.ibtauris.com

In the United States of America and Canada distributed by
Palgrave Macmillan a division of St. Martin's Press, 175 Fifth Avenue,
New York NY 10010

ISBN: 978 1 84511 538 8

A full CIP record for this book is available from the British Library
A full CIP record is available from the Library of Congress

Library of Congress Catalog Card Number: available

Designed and Typeset by 4word Ltd, Bristol, UK
Printed and bound by TJ International Ltd, Padstow, Cornwall, UK

CONTENTS

LIST OF ILLUSTRATIONS

LIST OF MAPS

ACKNOWLEDGEMENTS

All round the world we have had the support and encouragement of many senior figures in the interlocking worlds of politics, security and energy. Ministers, policymakers, senior government officials, expert analysts, energy company executives and operatives, academicians, think-tank gurus, from Oslo to Tokyo, from Doha to Moscow, from Berlin to New York, from Vienna to Beirut, have all helped us with unfailing courtesy and interest in our work. They are too numerous to mention individually, although several of them are referred to in the text.

However, back in the UK there are five particular groups of counsellors and wise guides to whom we owe a special debt of thanks. They are:

1) Our fellow members of the Windsor Energy Group, notably the steadily supportive Paul Tempest.
2) Our fellow members of the British Institute of Energy Economists, who, again, have opened many doors and been constantly helpful.
3) The executives and staff of Middle East Consulting (MEC) who have encouraged us all along the way to get this book written.

4) Colleagues in the Surrey Energy Economics Centre at the University of Surrey, and in particular Professor Lester Hunt.

5) A truly amazing galaxy of ambassadors and diplomats, both the British ones in overseas posts and the foreign ones based in the UK, many of whom have gone to quite extraordinary lengths to put us in contact with the key energy people from their countries. There are many, but we would pick for particularly warm thanks Bjarne Lindstrom, Norway's genial ambassador; Khalid Al-Mansouri, Qatar's most assiduous ambassador in London; Yoshiji Nogami, the Japanese ambassador, and his small army of counsellors, First, Second and Third Secretaries in the London Embassy; Jihad Mortada, the former Lebanese ambassador; Yury Fedotov, the Russian ambassador; and Khalid Al-Duweisan, the Kuwaiti ambassador in London for many years past and now doyen of them all.

All these amongst many others were always ready with facts, figures, contacts and entrées of every kind, although what we have made of them is entirely our responsibility.

Our thanks should also go to all the staff at I.B.Tauris, including our patient editor, Liz Friend-Smith, and the wise Dr. Lester Crook, for their friendly guidance and support in getting this book published in a formidably short timescale.

We put last what should perhaps have been first – our thanks to each other as co-authors. Our backgrounds could not be more different. Yet perhaps out of this difference a book was born. From one side has come the energetic determination to produce a work on this subject, the academic expertise in many energy-related fields and in Arab and Middle-Eastern ways of thinking, the unwavering enthusiasm and the determination to respond to the concerns of younger generations in many countries who look with profound unease on our troubled world and its linked energy and environmental dilemmas.

From the other side has come quite a few years of experience in government, politics, finance and international matters, plus the journalist's readiness to put the marvellous English language to work in seeking to untangle confusions and to slaughter a few of the sacred cows which graze in energy pastures.

Readers with an idle moment to spare may like to guess which of the authors made which of these contributions to the undertaking.

INTRODUCTION

Big changes are coming in the world energy scene. The impact will be felt on all our lives.

Yet curiously there is little public interest in this prospect. The overwhelming mood is that something will turn up to keep the lights on and the wheels turning. Energy issues do not make front-page news or top publishers' lists.

To be sure there is plenty of discussion – mostly in more prosperous circles – about global warming, climate change and the need for lower carbon emissions – and the government policies to encourage them.

But the timescale on climate matters is inevitably very long indeed. By contrast the problems of energy reliability and security are right on our doorstep, here and now.

If these immediate energy issues are handled right, so this book seeks to show, a pathway can be found through to a much brighter, greener future, in which the growth of greenhouse gases in the atmosphere can indeed be curbed and maybe climatic disasters averted.

But if they are handled wrong, if the focus is on the wrong priorities, then both goals – energy security now and climate security later – could be seriously jeopardized.

The scene is confusing and the timescales muddled. Of course, there have been oil shocks and energy shocks before, notably back in the 1970s and 1980s. But somehow things recovered. Today, these are some of the questions people almost everywhere are asking:

Are we just in the usual cycle of high oil prices and costly power bills, with lower prices and plentiful supplies to follow, or is this time something different?

Are the wars and tensions, especially in the Middle-East, really about oil and energy supplies?

Can consumers round the world, and especially in Western Europe, trust the Russians, with their huge oil and gas resources?

Should we all be going for hybrid (that is high-mileage combined electric and petrol-driven) cars and home wind-pylons on the roof? Is that the answer?

Are China, India and booming Asia going to drain away all the oil and gas we need in the West?

Will safe nuclear power and cleanly-burnt coal come back to save the day?

If we in the West bust ourselves to cut carbon emissions and save energy, will it make the slightest difference unless the rest of the world follows? Can there ever be a truly worldwide scheme to cut carbon?

And how do either energy problems or climate change issues affect the poorer countries of the world, with their desperate need for cheap and plentiful energy to develop?

This book takes the reader on a journey through these dilemmas and puzzles, although it is a journey of which the end is by no means in sight.

Yet our basic message is positive. The opportunities – often missed in the past – are still there to master the energy situation, preserve the best things in our planetary environment and guide the world through to calmer and less threatening waters.

But to do that we need new policies, new priorities and, above all, new and much more compelling messages to move both governments and people in the right direction. Neither the evangelism of environmentalists nor blind faith in the power of markets will see us through. The dangers of the more distant future must be woven together in the global mind with the imminent challenges of the present. How is that to be done? Please read on.

David Howell and Carole Nakhle
London and Guildford. New Year, 2007.

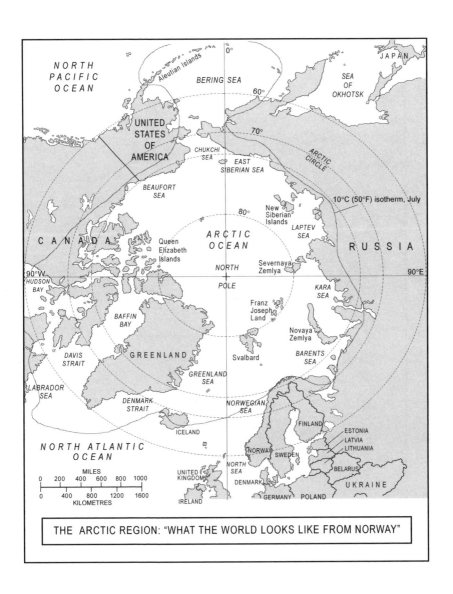

THE ARCTIC REGION: "WHAT THE WORLD LOOKS LIKE FROM NORWAY"

THE CAUCASUS
AND
CENTRAL ASIA

S I A

• Yekaterinburg

Omsk•

• Chelyabinsk
Petropaviosk•

• Qostanay

Paviodar•

Semey• • Öskemen

• Astana

• Qaraghandy

• Karamay

K A Z A K H S T A N

*LAKE
BALKHASH*

ARAL SEA Karamay•

Almatay•

Taras• • Bishkek • Aksu
Talas

• Nukus

UZBEKISTAN

Shymkent•

K Y R G Y Z S T A N

Tashkent•

Andijon•

C H I N A

• Kashi

• Urganch

Navoly• Jizzax•
Buxoro•

• Samarqand

TAJIKISTAN

I S T A N
Türkmenabat

• Qarshi • Dushanbe

Qürghoonteppa•

• Mary

Termiz•

PAKISTAN

ashhad•

Gusgy•

Kabul•

I N D I A

Islamabad•

A F G H A N I S T A N

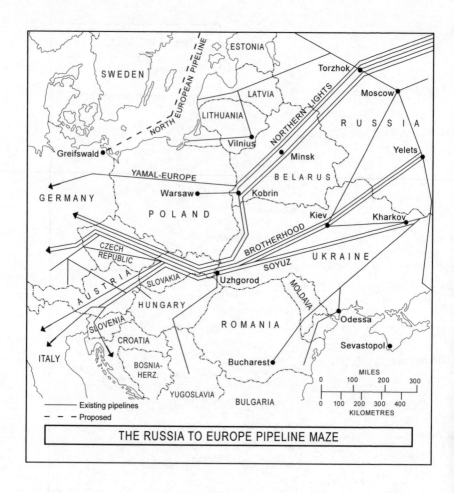

THE RUSSIA TO EUROPE PIPELINE MAZE

CHAPTER ONE

THE GRAND ALLIANCE

Can the future be brought into the present? How the searches for energy security and for climate security must be allied and combined to have impact and provide the escape map from the energy labyrinth of winding contradictions, conflicts and confusions. Right and wrong ways to a low-carbon future.

Everybody wants a green and pleasant future in which the world is saved from violent climate extremes, from floods, droughts, ever more terrible storms and other immense environmental and ecological catastrophes.

But they also want secure and reliable energy supplies, at reasonable cost, to guarantee uninterrupted light, warmth and comfort, to feed their industries and, in the case of the billions in poverty and on the edge of hunger, to lift them as quickly as possible to a better life. They want to travel, they want to fly, they want to prosper, they want to be free.

Can they have all these things? Can they have both climate security and energy security? Can they both have all the energy they need and yet avoid the environmental consequences of consuming so much? That is the question. To head off climate catastrophe it seems that there has to be a vast reduction in the burning of fossil fuels – that is oil, gas and coal, the chief culprits in pumping tonnes of carbon dioxide into the atmosphere and warming the planet much too fast.

But the current trends are all *strongly the other way* – and not just the trends but the most realistic and authoritative forecasts as well. Quantities of carbon in the atmosphere are rising worldwide and show no sign of falling. More oil is being consumed round the globe than ever before. Demand shows no sign of significant slackening and is far ahead of predictions, despite the doubling in the price of oil in recent years and high energy taxes in many countries on top of that.

Huge investments are being planned to transport and burn prodigious quantities of natural gas, especially in Northern Europe. Although cleaner than oil, gas still emits massive volumes of carbon. And the hunt for energy to fuel growth is turning the developing countries towards still more reliance on burning coal, of which India, China and America have about half the world's enormous reserves – enough to last for centuries. Yet it just happens that coal is the most carbon-intensive and the dirtiest of all energy resources – unless it can be cleaned and the massive carbon emissions captured and stored. But who is going to pay for that?

Former US Vice President Al Gore, amongst others, has set out the longer-term dangers with scarifying clarity, in his film and book *An Inconvenient Truth*. Mr Gore has now been appointed as Adviser to Gordon Brown, the UK Chancellor and about-to-be Prime Minister, on climate issues. His message, with its terrifying warnings of disaster unless carbon emissions are reduced, commands a wide audience. Alas, although his work is sub-titled 'A Planetary Emergency', it is weak on remedies to ease hardship, energy disruption and unrest in the next few years. It is becoming clear that there are other and more immediate inconvenient truths to face.

The first is that the struggle to control the climate – by far the biggest and most ambitious mission ever undertaken in human history – has to be totally global to have any effect. The

Oil takes off in the North Sea, 1980. The new Energy Secretary, David Howell, is briefed. Will it run out? No. Will it cost more and more? Yes.

biggest source now of greenhouse gases is America, but the biggest sources in the future will be China and India, with a third of the world's population. Even now, if the UK closed all its power stations, the carbon emissions saved would be the equivalent of a year's *increase* in emissions in China.

3

Experts and policy leaders go on nowadays about the global nature of this issue and that. But when it comes to climate the problem is truly, unchallengeably and inherently global. It cannot be anything else. The atmosphere does not sit in convenient puddles over regions of nation states. It swirls around the planet affecting everybody and being affected by everybody.

The second awkward reality is that, while appeals to the higher moral senses may have resonance amongst policymakers and the more comfortably off, the vast majority of humankind's choices have to be based on hard financial and economic considerations. The global catastrophe message is not nearly enough to persuade people to make sacrifices for their great grandchildren or change their lifestyles (or more frequently sheer survival 'styles'), or businesses to re-organize their energy intake, or governments round the world to change their policies radically. A much more compelling story has to be devised and much more powerful motives and incentives have to come into play.

A third stumbling block in the campaign to cut greenhouse gases is that the timescales are badly out of line. Energy needs are immediate, and the threats to world energy supply, whether from terrorism or political upheaval, are immediate. But actions to cut carbon have a hugely long lead time.

The much-discussed report by Sir Nicholas Stern, commissioned by the UK Government, on the economics of climate change, was quite candid about the timing issue. 'What we do now', it declared 'can only have a limited effect on the climate over the next 40 or 50 years'. In the second half of the twenty-first century all the efforts now to cut carbon emissions will, so scientists increasingly agree, have some benefit. They may prevent the tipping point at which the weather finally turns against humanity in a rage of destructive floods and freezes and boiling heatwaves.

The Stern Review had little new to add to this broad conclusion, but caused great excitement because it seemed to import hard-sounding economic calculations into a world of scientific predictions and, frankly, guesses. The feet of the British Prime Minister appeared to leave the ground when he described it as the most important report he had ever received. But in the end, despite Stern's brave efforts, the report was trying to give objective clothing to highly subjective issues. How many sacrifices we are prepared to make now to help future, and possibly much richer, generations is a matter of value judgments and personal feelings. Quantifying it is meaningless. Picking out some arbitrary discount value between costs and payments now and the possible benefits for distant generations hence (and Sir Nicholas picked a nice low one) stretches credulity. It depends, as with all economic modelling, on the assumptions with which one starts. In subsequent interviews Sir Nicholas seems acutely aware of this, unlike many of his report's excited interpreters.

Meanwhile, back in the real world, price, cheapness and reliability will be the deciders of the world energy mix. Somehow the future has to be brought into the present and the motives and fears which move people and nations *now* harnessed to the longer-term goals. Reducing the carbon in the atmosphere involves an intergenerational bargain and a huge leap of faith. But has anyone told the public that?

A fourth awkward truth is that the investment decisions required to transform the energy supply-and-demand patterns of the globe necessitate immensely long-term commitment, and therefore a high degree both of policy continuity and price predictability.

Yet not only do most governments come and go, with politicians inevitably living, and trying to survive, in the very near term, but the energy scene is fraught with extreme volatility. The oil price, which is the key to energy prices, soars and then slumps, taking the gas price with it and confounding capital investment calculations. When oil prices drop, as they well might in the future, everyone gives up on energy efficiency and energy saving and goes gratefully back to cheap oil (and gas and coal), thus undermining the climate warming struggle. They did so last time (in the 1980s when the oil price fell precipitately), and they could do so again.

Actually, that is not the whole story. America certainly went back to gas guzzlers (with some minor improvements in fuel economy) when oil prices plummeted after 1986, and soon lost interest in alternatives. Long-term plans for power investment were abandoned. Nuclear power went on hold and dependence on Middle-East oil, after falling in the previous decade, rose again higher than ever. Those who had invested heavily in renewables, such as wind power or solar power, lost their shirts.

In Japan, almost uniquely, energy-saving measures, and especially oil-saving measures, continued strongly all through the 1980s and 1990s, the years of cheap and easy energy supplies. The Japanese were driven by the old deep fear that they would be cut off from oil supplies, expensive or cheap. They were not going to be caught again.

But elsewhere, will the American-type scene be repeated in current circumstances as the world oil price dips then soars again? Are we doomed to an unending cycle of energy glut followed by renewed shortage and crisis, and all the while carbon-fed greenhouse gases mounting towards a total disaster point?

It ought to be possible to combine the urgent needs for energy security, as well as the urgent needs of the developing world, with the long-run fight against global warming. Together the twin goals of energy security and climate security ought to provide a truly motivating worldwide story which the prophecies of disaster some decades ahead lack the persuasive power to convey or turn into action. The requirement is long overdue to combine the mind-numbing technicalities and current dangers in the global energy supply pattern with the popular environmental concern about climate change, and this book tries to meet that challenge.

Harnessing these two causes – of energy security and climate security – would be to create a grand alliance, a grand unity, or at least harmony, of purpose which the world so conspicuously lacks at present. This would be *the* synthesis of argument and purpose to set the world firmly on the road to the energy transition, meaning the transition in all our lives and labours, which is still only hesitantly under way.

Three reasons are to hand which give hope that a repeat of the past is not inevitable.

First, oil consumers everywhere, in both the richer and the poorer worlds, are once bitten and twice shy. After the oil price collapse of the 1980s it all looked so easy. Oil was cheap and plentiful so why worry? Yet that proved a dangerous delusion. This time consumers and governments are more savvy – or so it is to be hoped. They may see that the pause in the relentless upward movement of oil prices may only be a temporary blip.

Second, the whole world energy situation looks far less secure overall. The Middle East, source of so much oil, is sinking into worse turbulence. Other oil-producing centres seem to be equally enveloped in political turmoil and instability. It becomes harder and harder to side with Dr. Pangloss and assume it is all going to correct itself and be for the best.

Third, the longer-term concerns about climate change are at last getting through. The timescale may seem eternal but the general worry about global warming, combined with visible and dramatic *current* signs of climate change, albeit now too late to reverse, have truly gripped the public mind.

If schemes for pricing carbon, and thereby presenting consumers with the true cost of the energy they consume, can somehow be established worldwide, and win confidence, then the process of real change could at last be triggered. But the 'ifs' are very big, the track record so far is not inspiring and it has to be remembered that governments and institutions which create carbon prices, via permits etc., are made up of mere humans who can change their minds or easily be replaced.

People who are struggling to meet existing energy bills and fuel costs do not take kindly to further painful charges and taxes – all in the name of meeting a seemingly distant danger. And politicians who want to stay in office are nervous of adding to the tax burden on energy when the reasons for the higher imposts are dimly, or not at all, understood.

In studies of climate change there is often talk of market failure, in that the full external costs of burning oil and other fuels do not always fall on the polluter and user. The danger with that kind of thinking is that it leads not to ways to mend market mechanisms – the only route that will work in practice – but to leaping in with more state controls and lofty institutional centralism.

Nonetheless, despite the pitfalls, these messages are the ones which should be at the top of government agendas and should ride together – that the world wants energy which is both secure *and* leads to lower carbon emissions. The error would be to rely on environmental fears alone, and on carbon reduction goals as *the* priority. The belief is noble but the science is still just too uncertain and people know that. Besides, who has time for worrying about carbon emissions when there is doubt about where their next meal is coming from?

The lead has to be given in terms of immediate energy security dangers and needs – which are considerable and have major implications for world growth and stability – as well as in longer-run climate terms. Otherwise, without real economic incentives to save energy and invest in cleaner alternatives, all the long-term hopes for climate change are bound to be undermined. Short- and medium-term crises will capsize longer-term hopes.

A Dose of Realism

It is noticeable that the biggest and most successful national endeavour to cut carbon emissions in modern times was driven by motives which had little or nothing to do with saving carbon. The huge French nuclear power programme, which has dotted France with 58 nuclear power stations, was started in the 1970s to reduce French dependence on imported oil and on coal. Today French electricity generation is virtually carbon-free. The path to a low carbon future can come from many different origins.

Scratch a populist politician or green-inclined celebrity guru and the chances are you will get an extended answer about long run global warming and the need for a low-carbon future and how to achieve it. For example, both the recent major energy strategy announcements by the British Government – its Energy Review in 2006 and its energy policy 'conclusion' in July of the same year – focus almost entirely on the long term.[1] Indeed the first of these two documents specifically rules out examination of short-term energy goals – which is just as well, because very few of them are being met.

Yet as Maynard Keynes remarked, 'in the long run we are all dead'.[2] It is in the short run or term that we all live and breathe. Anyhow, when the long term does arrive it somehow always turns out to be the short term all over again. Like the point where the rainbow ends, it is never there once you get to it. All that marks the spot is a mound of new documents and strategies to explain why the elusive long-term solutions to everything still lie over the next hill and why the search must continue with fresh determination. And so *ad infinitum.*

The dilemmas over energy are here and now and demand immediate and decisive action. Of course, it is right that we should be striving now to leave the next generation a less dangerous world than the previous generation has bequeathed to us. Climate change is real and full of dangers (as well as some benefits). Unfortunately, none of these numerous schemes to limit carbon, or the very long-range strategic insights, road maps, hopes and plans for more sustainable development, or sometimes blood-curdling glimpses of the future global landscape which accompany them, are of much direct help to the struggling billions trying to lift themselves at this moment out of paralysing poverty.

Even while you read this, the huge oil price increases of recent years, despite being below their recent peak, are murdering development and snuffing out the take-off attempts of the world's poorest countries. It has been estimated that the ten poorest countries of Africa in 2005 and 2006 have paid out more additional foreign exchange to meet price increases for oil and oil products (mostly gasoline) than the total amount they received in development aid and foreign direct investment. A $10 oil price increase knocks 1 per cent off the GDP of African oil importers. The attention paid to the immediate energy problems of developing countries is pitifully slight.

Nor will imposing visions or calls to mount a new climate change crusade help the poorest in the colder countries in the winters immediately ahead. None of these will do much to alleviate the much higher utility bills coming into every home or the greatly increased fuel hardship so caused, the doubled price of petrol (gasoline), entering into the cost of every journey and into all commercial transport costs – which means just about everything. None of them will give guidance to investors, or governments, about the right and urgently required decisions today to keep the lights on and homes warm tomorrow or about the right energy mix for the future.

Scientists argue rumbustiously with each other over many of these issues. But on one aspect they are broadly agreed. The timescale for halting, or even slowing down, the process of global warming is very, very long. Warriors for low carbon and worriers about immediate energy supply dangers should not therefore be at odds – they should be allies. Their agendas coincide even if their timescales do not.

In the meantime visible and sometimes violent changes in the world climate are already upon us. In the blunt and candid words of the UK Government's Panel on Climate Change: 'Even if greenhouse gases level off now, warming will continue at the current rate for several decades.'

Thanks in part to the activities of previous generations the level of in the atmosphere is already dangerously high, and shows no sign whatever of levelling off. An unstoppable process of melting ice in the Arctic region is under way, uncovering huge stores of methane and carbon dioxide (classic greenhouse gases) which, once they reach the atmosphere, will accelerate the warming affect still faster.

There has to be a dose of realism in all this. Changing the world's climate is going to take decades, even centuries, to produce results. Wise policymakers and advisers will approach the issue with humility and seek to move step by cautious step. They will avoid the excitable hype of columnists, populist politicians and even persuasive investment managers that decisions and measures taken now, today, can have a big current impact and miraculously influence weather pattern changes already in the pipeline.

To get the current position absolutely clear, and resting squarely on the facts, scientists have established that the world's atmosphere now contains 380 parts of greenhouse gases (carbon, methane, etc.) per million, having for the last half million years or so ranged between 180 and 280. That's already enough to disturb the weather pattern savagely – incidentally with major implications for near-term energy security, such as sunken oil-rigs and flooded mines. There is no hope of cutting the current carbon level, but is there any hope of checking further growth? That growth is now running at about two parts per million a year, so that by 2050 it could be (and here the 'facts' are getting a bit speculative) nearing 550 parts per million, which it is believed would trigger a surge of new climate catastrophes.

Maybe this further rise can be checked by burning less fossil fuel and penalizing many other carbon-emitting activities worldwide. Maybe it cannot. And maybe it will be much less costly to act now than later, as the Stern report argues. But what is crystal clear is that, starting from here and now, long-run targets for checking further carbon growth, however 'demanding', or backed by stern-sounding laws, stand not the slimmest hope of being achieved unless they are seen as necessary to protect people and industry against violent price movements, against power cuts and supply disruption, or against more oil shocks, as the Middle East and the Islamic world continue to boil. Nor will they do much to prepare the world for, or help it adjust to, the big climate changes which are already in the pipeline, *and which no amount of carbon curbing will now avert*. That will require expensive adaptation on a major scale to climatic developments which are now inevitable and unavoidable.

Calling targets 'demanding' is a favourite piece of government-speak. It means they have not been met so far and are unlikely to be met in the future. But repeating them, and constantly raising them, sounds good and purposeful. At present the world is adding seven billion tonnes of carbon annually to the atmosphere. For those, including these authors, who find it hard to imagine that volume, just accept that carbon emissions are rising very fast all the time, and will do so for years to come, whatever the rhetoricians say or promise.

What is most striking about the current world energy pattern is its extreme precariousness and the extreme uncertainty it creates. Thanks to the wonders of the microchip the energy supply chain is a far more integrated system in every way than it was at the time of the last great oil shocks 25 years ago. But it is considerably more vulnerable. Hugely integrated means hugely sensitive. The world is tip-toeing round the edge of chaos, an edge from which it could slip into a vortex of disorder, financial, economic and social, almost any morning.

Can today's political leaders be relied upon not to do anything too stupid to upset the balance? No – no more than they could be relied upon in the twentieth century. The march of folly is ready at any moment to roll on. The Arab Gulf states all warned of the dangers in attacking Iraq. They now live in daily fear of the consequences which might follow an American-led strike against Iran, including the firing of Iranian missiles into their shiny new high-rise hotels and offices – easily within range – and into their almost hopelessly vulnerable oil and gas installations.

An 'event' in any of these areas will send oil prices soaring again, financial markets crashing and investors scurrying for cover. Blocking the Straits of Hormuz could also be tempting. At the mouth of the Arabian Gulf, the Straits are 21 miles wide but the safe shipping lanes down the middle are much narrower. There is no difficulty about scattering these lanes with mines, at which point all insurance would be refused for ships navigating the stretch. This would stop about 18 million barrels of oil a day, or refined product equivalents, reaching world markets. That is about one fifth of world consumption. The result would be global chaos.

So can practical moves be made to ensure that short term energy needs and long-term low-carbon aims reinforce each other instead of being in conflict? Is there a way out, an escape route from the energy labyrinth? Definitely yes, say the chapters which follow. They suggest that if governments and their advisers think clearly, debate honestly and openly and put first things first, a great deal can be done to prepare for and handle the threats to energy security which lie immediately ahead.

The shapes of these challenges are already visible. No futurology is required to see the near-term threats or disruptions which will now flow from the super-fragile situation in world oil markets, where current spare capacity to pump extra oil – to cushion the inevitable continuing shocks to the oil supply chain – is now miniscule.

No futurology is needed to see that oil from Iraq, potentially one of the world's biggest oil sources, will remain a trickle for years to come; or that the aggressive stance of Iran, another oil giant, will cast a growing shadow over the scene;

or that the Kingdom of Saudi Arabia, the biggest oil and gas producer of all, will become more and more vulnerable to devastating and targeted terrorist attacks and extremist pressures; or that the next largest energy giant, Russia, will stay politically very unsettled; or that volatile governments will continue in Venezuela and Bolivia and a dozen other places; or that Nigeria, Africa's biggest oil producer, will face further pipeline sabotage. And no crystal-gazing is required to see the rising international tensions as nations scramble in competition for access to oil resources and control of the colossal wealth which flows from them.

The energy scene is famous for its long perspectives, its projects stretching years into the misty future. But to meet these challenges the need now is for leadership which unites both near-term dangers and long term hopes. And to get the world's policymakers to concentrate on the immediate, on the tripwires just ahead, the focus of energy security policy needs to be drastically shortened, not lengthened. This process will be greatly helped not just by a larger supply of candour and frankness on the part of political leaders about the present dangers and trends, but also by a better understanding of how we got into this mess and what very quick moves are needed and possible to get out of it.

Chapter Two seeks to explain how the apparently golden years of the 1980s and 1990s, when energy seemed cheap and plentiful and no one bothered too much about future problems, in fact concealed the seeds of future chaos and dangers which are now upon us with a vengeance. Have we learned these lessons or are we condemned to see them repeated one more time, or many more times?

Chapter Three sets the broad current energy scene and the immediate dangers. It seeks to correct many myths and illusions which are clouding current discussion and blanketing the landscape in a thick fog of contradictions and dubious statistics. It shows how extraordinarily hazardous and uncertain the global energy situation has been allowed to become, thanks to continuing and still growing world dependence on oil, and how exposed the poorest populations of the world now are. It explains how the UK Government, for one, has some of its priorities dangerously wrong, with current policies almost inextricably impaled on the horns of numerous dilemmas and key investment decisions about power supplies being paralysed. And it underscores the dismal reluctance in the salons of authority to face up to these dilemmas and 'inconvenient truths', prepare to meet near-term dangers or open the way to a better energy scene further ahead.

Chapters Four to Six dissect the whole energy supply scene, both conventional and unconventional, fossil-fuel-based and renewable, national and international. Chapter Four explains how misplaced some of the more common assumptions about energy supply really are, such as the frequently repeated dictum that the world is running out of oil. This is very far from being the case – or the problem. Scarcity of energy resources is not the issue, nor the reason for soaring prices. The problems lie not in the ground but in the surrounding geopolitics and in the machinery of production, transportation and final delivery to end-users, as well as in a fair amount of speculation and hoarding.

Amongst all the commodities oil is perhaps the one where both supply and demand are the most 'inelastic', as the economists like to call it, in the very short term. That is to say, when the price rises it takes quite a while for investors, companies and countries to find, invest in and open up new fields. And for reasons which will be set out later it is getting increasingly difficult and dangerous to do so.

Meanwhile, on the consumer side people just have to go on using oil and petrol for daily existence. Eventually they will adjust, when they realize they have to, but again it can take a long time.

On balance, oil will become, and probably stay, much more expensive than in the past, certainly less reliable and dirtier to burn than many other fuels. But at a price it will be there in plenty for decades to come.

Chapter Five focuses on the vast lattice-work of pipelines now pumping oil and gas across the planet, eastwards as well as westwards, and on the disputes, often violent, the dangers and vulnerabilities which the new pipeline politics generates.

Since Northern Europe's most crucial energy source for at least the next decade will be gas, the outcome of these machinations and manoeuvres will be of the profoundest significance.

Chapter Six shows what an immense range of alternative fuels, especially those from plant-derived carbohydrates and waste, and in the developing world from solar power, can be tapped speedily, given the right incentives, tax structures and other policies. It peers ahead to a possible world of low-carbon nuclear power generation which seems so obvious and yet raises so many difficulties. It tries to evaluate the possibilities as objectively as possible, weaving as honest a way as possible between the often over-enthusiastic lobbies for each energy source and their equally virulent debunkers.

Chapter Seven turns to the other side – the **demand** for energy and current consumption patterns and trends, as opposed to the supply sources. Increased energy efficiency is a wide open gateway. It is the route the low-carbon enthusiasts feel happiest to travel. No awkward questions about nuclear power if so much less energy can be engineered to produce more output.

But where are the triggers for change? The chapter sets out a full range of immediately achievable measures and actions which could have a quite dramatic impact in easing world energy tensions, as well as paving the way for substantial carbon curbs, if only policymakers would focus on them and if only those in authority, both in government and in business, would take the lead in explaining what can be done and how to harness market forces to new objectives. It details all the business moves, steps by government and manageable and practical changes at the level of the individual that are possible and achievable now to alleviate the fast-approaching upheavals.

Chapter Eight traces the winding and sometimes obstacle-strewn way forward, the necessary, urgent although complex route map for a manageable survival path and for an escape into calmer conditions in which immediate power needs can be met and long-term hopes and dreams can be pursued.

We can save energy in countless immediate ways, we can significantly curb the demand for oil, with all its costs and dangers, also in a reasonably short timescale, and we can invest promptly in a new and safer energy mix, given determination in all parts of our societies to do so. But that depends on our priorities and appreciation of the dangers. While eyes and minds focus on noble longer-term targets for the planet, we can all too easily fall into much more immediate traps.

One inch ahead is total darkness, goes the old saying. Even in the energy world with all its long views, its massive investment horizons and its love of scenarios, no one truly knows how much higher oil prices will rise, or how fast; the underlying trend is bound to be upwards, as the enormous costs and risks of new infrastructure and transport kick in. But oil is, after all, a commodity, although a highly politicized one, and commodity prices frequently zoom up and down. Some sources are confidently predicting oil above $100 a barrel by 2010 (Matt Simmons or Goldman Sachs), while others are talking of oil back at $40, or in a range hovering around $55 for the next ten years (for example, Henry Groppe Jr – who was so very right about the low oil price back in the 1980s).

Nor is anyone certain about what is going to happen from day to day on the geopolitical scene to disturb or destroy long-term ambitions for a more stable planet and for safer, cleaner and more secure energy flows in the distant future.

But in one respect the corner of the curtain can be lifted and what it reveals is danger – especially for the advanced societies of the West. If we lift it a little further we can see quite a distance into the near energy future – say, the next two or three years, which is time enough in most people's lives.

Over that sort of period the world faces a rough and painful ride, and moments of great risk and tension. The ride can be made less rough and less painful by more honest expositions on the part of those in authority, by clear-sighted policies replacing muddle and delusion, by a step change in international collaboration (the problems being both global and local), by bold business and market perceptions and decisions and by a skilled blending of local solutions and enlightened self-interest.

Much will depend on a far greater degree of public information and understanding about where we are, what can be done and how it will bring to pass a less uncomfortable – maybe even beneficial and profitable – phase in human affairs.

Conclusion

For the people of Western and Northern Europe – and probably elsewhere as well – serious shortfalls in electricity supply, brown-outs and power cuts and falls in gas pressure (and outright interruptions to industry), are now in prospect. Just ahead may well lie Russian failures to meet their gas supply undertakings, further petrol and diesel price increases, much higher electricity and gas prices (plus levies and taxes), still-rising emissions of carbon and other polluting gases into the world's atmosphere, much more burning of coal, especially in the developing world, ever increasing oil demands from China, rows about new nuclear power stations and about radioactive waste disposal, anger about forests of wind pylons in beauty spots, many icy winters ahead and nastier weather conditions everywhere.

Coming later on, say in ten years' time, if we are both lucky and wise, are plentiful clean, safe and reasonably priced energy supplies to power an energy-hungry world, a halt to the growth of carbon and other polluting gases in the atmosphere, cleanly processed coal, oil and gas in ample profusion, much more effi-cient use of energy supplies in homes and factories, more localized power generation, bioenergy at competitive prices, cheaper (and safer) nuclear power plants producing massive flows of carbon-free electricity with minimal waste safely handled, and maybe even calmer world weather conditions.

It is possible, but not if we stay on present paths. They lead only to a different, darker and poorer future.

The problem is to get from dangerous here to happier there. The future needs to be sharply different from the past in energy matters – and in much else besides. But those with authority, and with the power to give words wings, have yet to find the over-arching story with which to inspire, persuade and moti-vate. And without that today's hopes will be destroyed and tomorrow's fears grimly realized. We will remain trapped in the labyrinth. Here we begin the search for the way out.

CHAPTER TWO

THE PRICE OF
THE PAST

∽ ∽ ∽

In the previous phase of high energy tensions (in the
1980s) the world was bailed out by market forces –
lower oil demand along with higher oil supply. This time
there will no easy rescue. The dangers are bigger and
more complex and great difficulties lie just ahead for
people everywhere, in rich countries and in poor ones.

∽ ∽ ∽

VENICE – Wednesday 23 June 1980. Noon

In the high-ceilinged library of the Benedictine monastery on
the Isola di Giorgio Maggiore sit the leaders of the advanced
industrial nations, the G7, at a great round table. Officials
hover behind them.

Jimmy Carter is there, now in the last days of his
Presidency, having arrived in a warship moored in the lagoon
and said to have on board a staff of 800 advisers and security
men. President Valery Giscard d'Estaing of France is there,
together with Helmut Schmidt, the German Chancellor.

Margaret Thatcher is there, glittering and sharp and still only a year into her premiership. Brian Mulroney from Canada is there, facing an imminent electoral disaster. Saburo Okita is Japan's deeply experienced but rather silent representative. Francisco Cossiga, the Italian Prime Minister of the moment (in a seemingly rapid succession), makes up the seven.

Their united concern is energy. The whole Summit is dominated by energy worries. Oil prices have trebled in the recent months, sending violent judders through world financial markets and halting global economic growth. The leaders are joined this particular morning by energy and industrial ministers to decide what to do. Otto Lambsdorf, the German industry minister and dedicated free marketeer, is at the table; so is Francois Giraud, the architect of France's amazing programme of building no less than 40 new nuclear power stations. So is a David Howell, the UK Energy Minister and youngest member of the new Thatcher Cabinet.[3]

Papers in front of the Ministers on the table, including a draft communiqué, contain long lists of policy recommendations. They are all of the 'something must be done' variety, written to satisfy the politicians' impulses that action is demanded of them.

The main draft paper is on the edge of apocalyptic. It says that 'unless we cope with the energy problem we cannot cope with anything else'. There must be more oil production, more coal production, faster nuclear power station building and more conservation and energy efficiency – and if possible, 'dialogue' with OPEC, the fearsome oil-producers' cartel which seems to be holding the world to ransom.

Otto Lambsdorf shakes his head. No special intervention is needed because markets and events are already taking care of the situation. The growth of demand for oil round the world has already slowed, while new oil supplies are being opened up everywhere. Lower demand and higher output equals weaker oil prices. Greedy OPEC will shortly destroy the very markets by which it lives. That is obvious. The problems will resolve themselves.

Howell supports Lambsdorf but Carter and Giscard want something more detailed and positive. A communiqué is argued out with a bit of everything in it. The leaders rise, feeling they have done their best, although faintly uneasy that not all is well. They retire for lunch across the Lagoon.

Lambsdorf was of course right that morning. Even while the Venice Summit was taking place world oil prices were easing. The Americans, thoroughly frightened by the oil shocks of 1973–74 and again, even more painfully, of 1979–80, had already succeeded in reducing the famous oil-and-growth link dramatically. New binding laws on US car-makers were already biting and leading to big increases in fuel efficiency.[4] The old pattern, whereby every 1 per cent in growth of GDP had led to 0.8 per cent growth in oil consumption, had been bent right down to half that level. In other words oil consumption was still linked to growth, but much more weakly.

Most European countries, including the UK, were doing even better. No one was thinking much about China or India, or for that matter about terrorism or carbon emissions and global warming. Economic growth was anyway slowing everywhere. Within the coming decade oil prices would collapse entirely, slumping from a peak of $45, reached in 1980 (at the then current price), for a barrel of crude to $9 and even less by January 1986. Some cargoes were even being offered in the Gulf at $6 that month. The Middle East oil-producing states would see their revenues decimated. Oil Ministers like Sheikh Ahmed Zaki Yamani, once designated the most powerful man on earth, would be swept away as OPEC power withered. Cheap oil would be back – and stay back for 20 more years (barring a short-lived shock when Saddam Hussein invaded Kuwait in October 1990). The world could relax, lights could be left on, automobiles could get bigger again, nuclear power stations (always politically tricky) could be postponed. All was well.

But was it? Today, over a quarter of a century later, energy problems are back with a vengeance. And this time they are intensified and complicated by a raft of dangerous new issues and challenges. In the long sweep of history the 1980s turned out to be a false dawn. The perils of an oil-dependent world were submerged, not resolved, during the two decades after the Venice Summit gathering.

Oil at the Heart

Once again oil is at the heart of the problem, but the ripples go far wider. Crude oil prices have again been edging up to 1980 levels in real terms (before dipping, maybe to soar again), world oil consumption is growing faster than ever, and well ahead of expert predictions, and the balance between oil supply and demand is on a knife-edge, with surplus capacity almost non-existent.

A host of new questions and uncertainties hangs over the global energy scene. No one knows where the price will go next because the Middle East, the source of most of the world's supplies and the region with two-thirds of known oil reserves, remains in deeper political turmoil than ever. As long as spare production capacity is small every new event sends prices soaring again and speculation intensifying – whether it is a terrorist attack in Saudi Arabia, a new threat from Iran to mine the Straits of Hormuz,[5] another setback in Iraq, Israel's attack on Lebanon (despite neither country having any oil at all), or political lurches in Venezuela, or in Russia or in Nigeria or any of the other African oil-producing states. Per contra, a run of better news in the market sends prices sharply down, leaving the doomsters floundering and tempting the general public to shrug their shoulders and give up once more on energy economy.

In addition, there is a widespread and respected view – although not universally accepted – that the world is now running out of oil, meaning that new reserves have to be found and exploited in more and more remote and dangerous regions at higher and higher cost. The basic proposition of those who argue this way is that there is a fixed amount of oil and gas in the ground, almost all of it has been found and it is fast being depleted.

For example, in 2005 around 31 billion barrels of oil were consumed worldwide while only nine billion of new reserves were discovered (according to the not-very-reliable statistics – of which more presently). World oil production, goes the thesis, will therefore decline shortly. The world is already at, or is about to reach, so it is claimed, the famous 'Hubbert's Peak'. This high point, based on the calculations of a Shell geologist,

M. King Hubbert, in the 1960s, is defined as being a certain moment at which world discoveries in a given year, plus annual production of crude oil, fall behind world consumption in that year.

Have we reached this point? Nobody has a clue but new discoveries keep being made, leaving a very large question mark over the whole Hubbert prediction (more of this contentious issue presently).

Definitions in the oil world have anyway to be used very carefully and much argument between economists and experts swirls around not just this thesis but all the production figures which various oil-producers serve up.

Not least it turns out that new reserves of oil and gas have been increasing, not decreasing, in recent years. It all depends on technology and the oil price. As the technology for hunting for, finding and extracting new oil develops, and as the price makes spending on new oilfields more and more attractive, it obviously follows that more and more oil is 'found' and worth extracting, despite the increasingly hostile conditions. In the Gulf of Mexico some technicians, attempting to drill a mile under the seabed, are having to work in temperatures of 100°F. So 'proven reserves' estimates keep on being revised upwards.

It could indeed be that, for whatever reason, oil production for the time being has indeed 'peaked' from the world's already developed, low-cost and giant oilfields (mostly in the Middle East) and the proposition that 'we are using more than we are finding' holds true in that narrow sense. It is certainly true if applied to specific regions. The United States now has about 3 per cent of the world's proven oil reserves and drinks about 25 per cent of world output (with 4.6 per cent of the world's population). Mr Hubbert rose to fame by predicting that the US situation would peak in 1975, and he was just about right. But his peaking dates for the world as a whole keep retreating into the future. Even the Hubbert predictions for the United States could now be invalidated by Chevron's huge new reported find around 175 miles offshore from Louisiana. If Chevron's optimism is soundly based, this alone could add 50 per cent to the entire oil and gas reserves of the USA.

The same sort of argument could be applied to the UK section of the North Sea, where output has indeed 'peaked' in straightforward annual production terms, both for oil and for gas. Annual UK oil and gas consumption is now higher than production from the UK Continental Shelf, and the UK is again, after about 20 years, a net importer of oil and gas on a growing scale.

This should be conditioning UK foreign and energy policy, and driving the search for diversity in energy supply sources harder than ever. But does this prove that the UK is therefore 'running out' of oil and gas? No, because the North Sea was from the start an international province where the investing companies had the right to sell their product anywhere they wished (otherwise they might not have invested all the necessary billions in the first place). The concept of self-sufficiency – i.e. equating North Sea UK production with UK consumption in any one year – was always a paper exercise. The UK is an inextricable part of a huge global system of supply and demand as, for that matter, are America and China, although some Chinese planners do not seem to realize it. As will be explained, the Americans cannot drill their way out of their oil problems, nor can the Chinese bargain, promise and buy their way out.

Have the Oil Giants Had Their Day?

As for the great multinational oil companies, once sailing serenely on a sea of cheap and accessible oil, they now find themselves shut out of new oilfields. Shell has been ordered to stop building the pipelines it needs for its Sakhalin-2 project and has now been compelled, under environmental pressures from the Russian authorities, to sell out the main interest in Sakhalin-2 to Gazprom.

BP's hopes for plentiful new oil access in Eastern Siberia are looking less bright, and for 2006 it reported an actual fall in production, while of course meanwhile maintaining very high profits – a harbinger of things to come, and come soon.

In many places state oil companies are pushing out the old international companies, successors to the once dominant seven sister giants of the oil world.[6]

23

Aramco, the totally state-owned Saudi Arabian oil company, now has ten times the reserves on its books of those available to ExxonMobil, the largest remaining independent oil company. Venezuelan and Bolivian leaders, buoyed up by mounting oil revenues, have seized the oil and gas assets of American and other foreign companies.

Russia is pursuing a sort of grandmother's footsteps policy on intrusion by the state into the huge private oil and gas sector. Amidst protestations that it wants a healthy private and foreign investment sector, it is steadily moving the other way. Yukos has been swallowed up by the Russian state on various political pretexts, and its chairman, Mikhail Khodorkovsky, tumbled into jail. Foreign investors, notably BP and Shell, have been shut out of the most promising new exploration regions. New limits on foreign-owned companies have been introduced. And so the gradual creep continues.

It is not just a question of state-owned oil companies gripping more tightly their own national oil resources. These national concerns, driven more by political agendas than by shareholder interests, are branching out increasingly onto the international scene. Round the world private sector oil companies find themselves outbid for new oil concessions by thinly disguised government agencies with no shareholders and pots of money.

Firms like Sinopec Corp. or China National Oil Corporation or CNOOC, with seemingly unlimited cash to spend, and the Chinese Government close behind them, are the new masters of the oil universe as they mop up concessions and access rights from Angola to Nigeria to Venezuela and from Canada to Iran. Gazprom, the Russian state monopoly, is busy in talks with a dozen oil and gas concerns round the world. Its increasingly specific agreements with Sonatrach, the Algerian state oil and gas business, have raised fears about a new 'Gas OPEC' to squeeze over-dependent European consumers. (See Chapter Four.)

Meanwhile the global appetite for oil races ahead, still dominated mostly by America's colossal imports but now supplemented by a growing Chinese oil thirst,[7] with India some way behind but on the same upward curve. Earlier in the year of that Venice Summit (1980) the Chinese had been in London in force, in a delegation led by their then top man, Hua GwoFeng, Chairman of the Praesidium. Then the talk had all been about China as a great oil producer. Offshore developments were going to supply all the oil China needed and more. The Chinese Energy Minister slapped David Howell on the back and promised he would be invited as the first oil gushed out of their newest offshore platform.

But there was no great offshore oil flow and most of the new wells proved disappointing. China, which was supposed to be the great oil producer, is now back as the great importer, still far behind the USA (at four million barrels a day of imports against America's 13 million) but clearly heading fast upwards.[8] Chairman Hua was bundled out soon after his London visit. No invitation to see the gushing Chinese oil ever came.

Global consumption is now just on 1,000 barrels of oil every second, or one full Olympic-size swimming pool of oil every 15 seconds,[9] far more than most experts and authorities, or the industry itself, ever expected or predicted. There is much political speech-making – with President George Bush talking about America's 'oil addiction' – but little sign of withdrawal or of governments being willing to inflict the pain that inevitably goes with it.

Prepare for Turbulence

The world is therefore again in for new energy shocks in the years immediately ahead. Of course it is always easier for policymakers to talk about longer-term issues and solutions, and of course the greatest longer-term issue, the challenge of rising greenhouse gases and global warming, must be faced. But the threats to energy security, and to the health and stability of our energy-dependent societies, are immediate. The longer-term battle to curb the further growth of world carbon emissions (so far failing miserably because only selectively supported) must be fought for the sake of future generations and the survival of the planet. As noted earlier, world carbon emissions are still soaring – currently by seven billion tonnes a year from energy use, with much more added from agriculture. But then oceans and plants absorb several billions as well. So one has to be careful about net figures. As always in energy matters the statistics are highly pliable.

But unless the shorter-term crises, now building up fast, are also handled and managed effectively, with willing popular understanding and support, there will be no political will or resources left to cope with carbon emissions or much else. Hopes for a low-carbon future will be overwhelmed by even bigger challenges to daily life and survival.

There is no single obvious answer, no magic bullet this time to meet the immediate difficulties. Last time – following the oil shocks of the 1970s and early 1980s – the energy cycle bailed the world out. As so often glut followed scarcity. Very large and sudden price rises worked their market spell in the way Lambsdorf (and others) predicted. New oil and gas fields were identified, drilled and developed in response to the high price prospect. Economies in the use of oil expanded. With supply up and demand down relatively cheap and plentiful supplies of oil and gas naturally returned.

But this time the energy cycle may not save the day, nor would that be a bad thing. Years of weak oil prices may have been a joy for the consumer and for the oil-drinking advanced world. But the inevitable price paid on the supply side has been weak investment in the entire supply chain, from exploration and development through to production and refinery processing, and through to every kind of equipment supply in the chain in between. These were years in which no one wanted to spend too much on new rigs, new platforms, new drilling equipment or new tankers. Engineers moved away to other industries. Students turned to trendier subjects like global warming and renewable fuel sources.[10]

Not only did the sudden collapse of oil prices at the end of 1985 and the beginning of 1986 destroy incentives in the years following to spend big money on exploration and drilling for new reserves and for developing fields in areas which were both remote and dangerous, at the same time investment in up-to-date refinery capacity ceased altogether. Today that price, too, is being paid in outdated refineries (the last new one in the U.S.A. was built in 1976, although the Iranians are reported to be building new ones now) and in a serious mismatch between the kind of crude oil existing refineries can handle (mostly the light or so-called 'sweet' variety) and the heavy sticky stuff the giant oil wells of Saudi Arabia and the Gulf states now produce.

This time the new factor is massive and seemingly unfaltering extra oil demand, especially from the rising Asian powers. Much higher demand has collided head-on with a strained and unprepared supply chain. As the pendulum has swung violently from cheap and plentiful to expensive and scarce, shortages and strains have rapidly developed. And as the oil price has climbed a new investment rush has gathered force, scooping up every available oil rig and platform the world over. This in turn has sharply increased exploration costs.

A meeting in a Church

There was another avenue which looked open in the 1980s but led nowhere and still remains closed today. The world leaders at Venice in 1980 believed, correctly at that time, that OPEC could produce a lot more oil (in contrast to the tight position today). More contacts and dialogue with OPEC were urged. The UK was, by rotation, in the chair that year of the International Energy Agency – the 'club' of the main oil-consuming nations. The UK was also becoming a nation with a foot in both consumer and producer camps, as North Sea oil and gas output rose fast. British Ministers were therefore expected to take the initiative in developing this dialogue between supplier and customer interests.

This involved both visits by the then UK Energy Secretary to Saudi Arabia (and the other main Gulf producers) and a series of private meetings in London with the main OPEC movers and shakers, who at that stage (1980) were, in particular, Sheikh Ahmed Zaki Yamani, Saudi Arabia's oil minister, who seemed to have become OPEC's chief spokesman, and Sheik Ali Khalifa Al-Sabah, oil minister of Kuwait.

The meetings took place, rather oddly, in a church, or rather a former church, converted to secular pleasures, just off London's Belgrave Square. Now flourishing as the famous restaurant Mosimanns, it was in the early 1980s a club called The Belfry, owned by an acquaintance of Sheikh Yamani. Discussions took place literally in the former belfry room up several flights of stairs.

The talks were pleasant (as always with Zaki Yamani, a man of huge charm) but fruitless. Behind them was the vague and unspoken thought that if OPEC would keep the taps reasonably wide open and ensure that crude oil prices stayed in, say, the high teens and went no higher (in dollars), the consuming countries would prove good customers and not pile on too many taxes at the consumption end. This would satisfy, so the theory went, the endless OPEC grumblings that the consuming countries, especially the Europeans, were collecting all the benefit of high prices in 'rent', so why should OPEC engineer lower oil prices while the consumer nations maintained high prices through local taxes and collected all the surplus?

But Sheikh Yamani, while duly and predictably making this point, remained adamant that OPEC would continue to restrict output tightly and issued repeated warnings about the need for careful depletion of this precious resource under the ground. If there were obvious dangers that ever higher oil prices would eventually lead to a halt in world growth and a collapse of oil markets, he did not at that point show awareness of them – although that is what indeed happened in due course. The greater concern of both oil ministers seemed to be to press the UK, with its rising North Sea oil production, to join the cartel and become a member of OPEC. In this, of course, the Thatcher cabinet had zero interest. The very idea was anathema. In the climate that prevailed there was no basis for any kind of 'deal' between producers and consumers to stabilize oil prices and nothing was achieved. Other forces and events would do the work instead.

(Perhaps with hindsight these talks did have some eventual impact because, after the price collapse of 1986 and Yamani's departure, OPEC tried to adjust production so as to keep the oil price between $22 and $28 a barrel. Again in 2000, as

Down goes a giant oil tanker – a grim reminder of the not-so-hidden costs of oil.

prices first hit rock bottom once more, at $10 a barrel (in 1999), and then began to take off, driven by climbing demand and supply increasingly restrained by lack of new investment, and by Middle-East war and terror, OPEC started talking about minimum price floors and price ceilings. First it was $30 dollars that would be the target price, then $40. Then as demand, oil trader speculation and market tightness drove the price higher still, all efforts at control were abandoned.)

No Cushions This Time

Today things seem back where they started, with oil prices again hovering near 1980 heights, faltering then climbing again, and the same worldwide symptoms of stagflation (prices rising but demand and investment falling) which shook the world economy in the early 1980s.

But why should price relief not come this time, as it did before, although with other painful side-effects, from slower world growth and generally weakened oil demand? The answer lies in the simplest economics of supply and demand. Supply is costing more than before and demand is rising faster than before.

When easily accessible oil, with very low production costs (mostly in the Middle East), was plentiful, the final price was low as well. Neither more expensive and remote oil regions, nor so-called unconventional oils (for instance from tar sands and bitumen lakes), could begin to compete.

Now that this 'low-cost oil' is running down, and now that the cheap oil regions themselves are anyway becoming much more dangerous, and therefore riskier to invest in (and thus not so low-cost after all), a trend to higher prices is unavoidable. Oil may be plentiful in places such as the Arctic region and Eastern Siberia, but trillions of dollars will be needed to get the infrastructure into place and to get it to the consumer.

There may be short-term blips as supply runs ahead for a moment, or demand suddenly sags. There could even be an increase in the spare capacity margin for a while. It is in the nature of world oil markets that they violently overshoot, both up and down. This is not just the work of sinister 'speculators' as some like to claim, although speculative froth certainly adds a few dollars to the price at times. When the oil price falls, textbook theory tells us that less oil ought to be forthcoming. But in practice, since most oil comes from countries with governments desperate for revenue (to meet lavish political promises and the demands of greedy intermediaries), the cry goes up for more oil to be pushed onto the market, so prices fall further still.

But it will be short-lived. Basically cheap oil has gone. The costs of recovery and marketing have become very substantially higher, so oil prices have moved on to a much higher plane. Like the humble oyster, once the cheapest of foods and now an expensive luxury, or a fish like cod, which used to be the fare of school dinners and is now a restaurant delicacy, oil will stay expensive, and wiser heads will plan accordingly. There is no escaping the harsh facts.

Escape Routes and Dead Ends

Expensive oil means expensive gas – at least in the short term as demand from homes, from factories and from power stations switches away from oil – which means expensive electricity, especially in the UK, where no less than 39 per cent of electricity supply comes on a normal day from gas turbines. (In 1980 it was 1 per cent.) So why cannot nuclear power fill the gap, as well as meeting carbon curbing goals, since while there may be plenty of carbon emissions in the process of constructing nuclear power stations when it comes to actual generation there are none?

The problem is that while expansion of nuclear power for electricity generation will help reduce carbon emissions in the distant future, it will take eight to ten years even to replace existing expiring plants, let alone expand nuclear electric output overall. And this assumes that government approval for new nuclear power construction, plus legal devices for 'speeding up' planning procedures, will somehow make it happen and worries about handling the intensely poisonous radio active waste will be laid to rest. Finland, which made the brave decision to go ahead with new nuclear stations, mainly to escape reliance on Russian nuclear-generated electricity, found that it took nine years to get the first one into operation. Huge construction delays have enveloped further plants.

The cost of building nuclear power stations is also a big deterrent. They may well prove impossible to finance, threatening to produce electricity only at prices far above their competitors such as gas and clean coal, and therefore turning off investors completely. Since the construction timescale is so immensely long, nobody really knows how things will look on start-up day. It may be that the most recent innovations in nuclear plant design, with serial building of identical plants, can bring both costs and construction time down somewhat. But the implication that new nuclear power will solve near-term energy problems, as appears to be at the heart of UK energy policy, is profoundly misleading. The timescales are completely wrong.

31

In the longer term a big expansion of civil nuclear power round the world may be waiting to happen. But long before the first new nuclear plant is ready there could be crippling gaps in the power supplies of several industrialized countries, including the UK, unless other more urgent decisions are made and carried out.

Other fossil fuel substitutes could help more immediately. Top of the list comes the old favourite, coal, which still dominates electricity generation in America, in China and in India. Coal is cheap, but coal is dirty. If somehow it can be 'cleaned' and the heavy carbon emissions captured and stored in deep-sea caverns, then with oil prices staying high (some say above $40), coal's comeback could be enormous. High hopes are placed by clean energy planners on the technology of carbon 'sequestration' – that is sucking out the carbon from coal as it is burned and piping it away to be buried under the sea, and possibly to be used to help pump more oil out of tired oil wells. The authors of recent EU pronouncements on European energy policy set much store by this – Europe fortunately still having plenty of coal.

There is just one large snag – which is that the whole technology is as yet unproven. It may work, it may not. It may cost the earth or the costs may be manageable. Nobody yet knows. (See Chapter Five for more on this.)

An alternative is underground gasification of coal – a well-tried technology which is being opened up again. But here, too, the cost figures are crucial. Oil at $70, with gas prices following on behind, leaves the field open for coal to be turned to gas underground, cleaned and made sulphur free, and then also carbon free, the carbon being piped away and stored underground or under the sea – and all at a nice, competitive price. But lower oil and gas prices reverse the economics, leaving those poised to invest big money in these new technologies uncertain and distinctly queasy.

Gas is the other fossil fuel which could stay at the centre of the energy mix, supplied both by pipeline and by ship in frozen form (Liquified Natural Gas, or LNG).[11] Those who want to escape all fossil fuel burning and leap direct to a carbon-free future do not like the idea of a big increase in gas dependence any more than increased coal burning, even though natural gas is a much cleaner fuel than either (its carbon emissions are 40 per cent lower than oil). But in Europe, at least, gas is going to dominate increasingly between now (2007) and 2020.

A critical point of confusion lies here in UK official minds between the politicized longing to put carbon reductions first, and the realization that the next few years will be all about coal and gas replacing oil, and that a big increase in gas network investment (including larger storage facilities) is a must. Trying to bypass or skirt round the coming coal and gas phases is like removing several rungs from a ladder. Gas is already well on the way to replacing oil for industrial processes and domestic heating, and across most of Europe has replaced oil completely for power generation.

But here, too, there are new and confusing questions, and plenty of risks. Continental Western Europe has drifted into 50 per cent dependency on Russian pipeline gas, and is heading for 60 per cent reliance by 2010. Austria already depends 90 per cent on Russian gas, Hungary and Poland almost 100 per cent. How reliable is the Russian supplier, and will the pipeline gas price continue to track the crude oil price? How reliable is the frozen alternative, which is being increasingly shipped from Algeria and from Qatar?

For the UK there is a particular problem with gas. Used to the luxury of relying on plentiful North Sea gas resources, the UK has not bothered too much about storage these last 20 years or so. But the gas splurge of the 1980s and 1990s, with new gas turbines being built all round the place, caught all the planners, and their political masters, badly short. Suddenly the UK is a big gas importer again and suddenly it needs big storage facilities to hold gas for weeks rather than days. Locating suitable underground sites for this kind of storage is a highly controversial business. No one wants one nearby.

Then there are the greener, supposedly entirely carbon-free alternatives. Can they be relied upon? These will undoubtedly help at the margin in years to come, and sooner still if the right policy steps are taken with promptness and vigour. On present trends by 2020 at least 10 per cent of Western Europe's energy supplies could come from alternatives and from green (renewable) sources – mainly hydroelectric power and wind farms. Solar power and biofuels will also play an increasing part. But the cost problem is the real headache. Costs are formidable (and partly oil-related), while the infrastructure needed to process and distribute plant-based fuels widely will require very large new investment. How do you hedge against the oil price sagging and making all your investment unprofitable – unless of course you can persuade governments and ministers to turn on the tap and provide fat subsidies to make sure you are always in pocket. The wind farm fraternity seem to have done just that, which is pleasant for them but leaves the European countryside littered with giant pylons producing electricity that are highly inefficient and probably always will be.

Next, there are all the plant-based oil sources. Later on, in Chapter Five, there is a wealth of detail about these alternative oil sources, for those who feel like it. Meanwhile, their development is generally on a far too leisurely a timescale to offset the tensions and pressures which have already arrived. The biofuels element in total petrol and diesel supply is still tiny – in the EU less than 1 per cent, in the USA rather more. But the whole transition would need to be vastly speeded up to make a big inroad into conventional oil consumption. With the right policies biomass fuels could in due course play a decisive part in meeting the global energy crisis immediately ahead. But it needs to be the right sort of biomass process. Africa today relies 80 per cent on biomass fuel but access to electricity, which is what its poorest communities really need, is minimal. The example of Brazil, where conventional oil imports have been almost phased out and ethanol from sugarcane is the major fuel for motorists, ought to have made energy planners pause and re-think long ago. Environmentalists rightly worry that rain forests are being cleared to make way for sugarcane and soya crops. But that should not blind them to the biofuel potential, if properly handled.

At least some of these alternatives are very promising. But because 20 years have been wasted assuming that cheap oil had returned and would stay, their development as decisive relief to power supplies and to tight energy prices may now be too late.

Fifteen years ago the world should have started investing heavily in alternative energy strategies. But it didn't. There seemed no need, there were no price signals to encourage manufacturers and investors that way and the political leaders, who ought to have stood a little taller and seen a little further ahead, failed notably to do so.

While idealists raise hopes about an oil-free and indeed fossil-fuel free future, and while scientists and policymakers focus on carbon pricing, greenhouse gases, rising sea levels and global warming over the next century, some deeply dangerous energy issues are already lapping at the door. Sir David King, the greatly respected chief scientific adviser to the British Government, while recognizing, as most people must, the seriousness of the longer-term issue of carbon emissions, has had the candour and courage to admit that for the next 30 or 40 years the damage has already been done.

It is the CO_2 already in the atmosphere, put there by previous generations, and already warming the ocean's surface, which is altering the climate and guaranteeing more extreme weather conditions, hurricanes, floods and dramatic seasonal changes. Nothing done now, he bravely admits, can make any difference to what the world faces in the next few years. In fact the violent weather conditions, generating out-of-season typhoons and exceptionally powerful hurricanes, such as Katrina and Rita, in the oil-producing Gulf of Mexico, are helping to intensify the short-term crisis by tipping platforms over and rupturing pipelines.

Those who put all their faith in, or urge others to put their faith in, the Kyoto Treaty, or Protocol, setting worldwide targets for cutting carbon emissions by 2008–12,[12] are going to be disappointed. First Kyoto will make little impact. India and China, comprising one third of all humanity, have not signed up to it, and the USA has a different agenda. Even if they had all been committed to the Kyoto emission targets, the impact on global warming 100 years from now would be negligible, and totally overwhelmed by other factors. It is a start, says its apologists, but is it even that? And could they be better starts in other more fruitful directions – a question which is posed uncomfortably by figures confirming that the maligned USA,

Chernobyl – the grim aftermath. But will future nuclear power stations be safer, smaller, cheaper?

non-signatory to Kyoto, has been reducing its growth of carbon emissions in recent years substantially, while in the European Union, despite all its emission controls, the rate of growth has been increasing![13]

Second, more violent climate extremes are already with us. CO_2 pumped into the atmosphere by past activity is already doing its unstoppable work. Reducing emissions now is a desirable aim for the benefit of future generations, but for most of those presently alive the urgent need is to prepare for and adapt to these extremes.

Third, the energy security dangers are now present and immediate. Kyoto in its current form will do nothing to meet these. Establishing a more reliable and secure energy supply system and reducing carbon emissions lie in part on the same road, which is a happy coincidence. But long before we get any results from cutting carbon emissions the world will probably be shaken by problems of security and supply disruption that could blow the highest hopes off course.

International planners are turning their attention to 'Beyond Kyoto' ideas for carbon reduction, and rightly so in the sense that the Kyoto arrangements are ineffectual and need bypassing. But minds should also be on 'Before Kyoto', that is to say on today's fragile energy supply scene.

While a low-carbon future is a vital goal, and work towards it must continue (somehow bringing on board the developing nations like China, without which all efforts elsewhere will be negated), the very much greater and more urgent priority than this super-long-range target is to address the here-and-now problems of painful further increases in energy prices, massive energy hardship, especially in poorer countries and communities, rising tensions between (and within) states and the major interruptions to industry, to transport and mobility and everyday life which lie just ahead – are indeed already with us.

Let's not sound too much like disaster-mongers, although chilling predictions of catastrophe and chaos always play quite well. A range of measures can be mobilized to meet and manage these threats, both on the supply and the demand sides. And, as always, amidst the crisis there are huge commercial opportunities for new energy-efficient products and services. Quite soon there will come a stage when at all levels of human society, from the humblest local and individual to the loftiest international, the realization will break through that there has to be urgent and big changes in the way we live and work. These will be driven less by longer run ideals (although they will help a little) than by hard economics and jarring threats to current practices. These could occur quite suddenly, long before the media or the policymakers catch up with what is happening. They may be occurring already in the transport field as the demand in the USA for hybrid, high-miles-per-gallon vehicles sweeps the market and waiting lists lengthen for popular hybrid products (in California the waiting time for new Toyota Prius hybrid automobiles is already two years).

Great fortunes will be made by those who see most clearly what is coming and how to meet new consumer needs and demands. New industries and industrial clusters will be spawned and grow, generating massive new employment opportunities as they do. Entirely new technologies, especially in the materials field, will emerge, alongside quite simple improvements in energy consuming items which will curb all forms of energy demand. Credit will accrue to those political leaders and policy-makers who speak frankly and realistically and give a prompt lead at governmental level in shaping the new framework within which markets and enterprise can deliver.

After 9/11 – Five Years of Folly

Meanwhile, developments on the world energy scene are both shaping and being shaped by a fast-changing international context. America's drift over the last two decades into ever greater reliance on oil, and on oil imports, has warped and distorted its foreign policy, with unhappy results.[14] Convinced that it must somehow stabilize and safeguard world oil markets and the main oil-producing regions, the USA has become more entangled in the Middle-East maze than ever before.

The 9/11 horror, the 'war' declared on terrorism and the generalized hunt for terrorists apparently gave the final push to the invasion of Iraq. But all along the Middle East, with its oil reserves, has been the focus of American concern, with American troops first in the Lebanon in the 1980s – with catastrophic results – then in Saudi Arabia to repel Saddam's Kuwait invasion in 1990, and now in and around Iraq itself. It was the Carter doctrine, (emanating from the President who, ironically, since leaving office has been a constant critic of American Middle-East policy) which stated in 1980 that 'any attempt by an outside force to gain control of the Persian Gulf will be regarded as an assault on the vital interests of the USA and... will be repelled by any means necessary, *including military force*'.

This is another reminder that the oil market is completely global. Add or subtract a barrel of oil anywhere and the whole world is affected. In practice the USA gets the bulk of its imported oil (12 million barrels a day) not from Middle East sources at all, but from Mexico and Venezuela (both politically highly unsettled) and from Africa. Only about a quarter of America's imports come from the Arabian Gulf region. But that makes no difference. An interruption anywhere in supply into the world pool of oil, or a big leap in consumption in any area (as now throughout Asia), has the same effect in creating a violent price spike, or lifting prices on to a higher plateau generally.

The American dream of a return to 'oil independence', or self-sufficiency, is just that. Not only is the world's energy system one single market of vast size, complexity and sensitivity; a dense web of new supply chain linkages, both physical and electronic, now makes the energy infrastructure of the globe infinitely more vulnerable to attack at its key points.

These attacks are occurring almost every day. Saudi Arabia, the central pillar of the oil-producing world, holding a quarter of the world's known oil reserves,[15] has had to face repeated attacks on its major oil installations. Two-thirds of its colossal production passes through two processing terminals, the larger one of which was attacked by Islamist extremists in 2002, and again in 2006. They were repelled but there will be more such attempts. These vital links in a closely integrated world system have been specifically targeted by Al Qaeda and identified as 'the umbilical chord and lifeline of the crusader community'.

In other oil-producing areas outside the Middle East the situation is little better. In an influential and profoundly expert study of the world energy crisis, and the dangers of continued oil dependency, published in the USA but little noticed in Britain, Professor Amory Lovins[16] and his colleagues point out that oil facilities are under frequent attack in Iraq, Colombia, Ecuador, Nigeria and Russia. Nigeria seems to be suffering an especially intense and frequent amount of pipeline sabotage. The top eight oil producers outside the Gulf area are all, without exception, in a politically unstable state.[17] Perhaps, he observes, it is the very fact that they produce oil, and receive a growing flow of petrodollars, which generates the destabilizing discontents and inequalities in these countries – the 'curse of oil' in visible operation.

The Hidden Costs of Oil

Those who depend on oil find they also have to defend its sources.

The extreme military cost to the West, and especially to the USA, of efforts to safeguard the world's oil heart and arteries adds enormously to the true 'cost' of oil. These are part of the 'hidden' costs of oil, in addition to the unquantified environmental costs, which could together add up to as much again as the visible market price which consumers pay.[18]

America's ever-deepening military involvement in the region, a reluctance or inability to settle the Israel-Palestine issue and the deeply flawed strategic impulse that democracy can be spread by overwhelming force, have helped to impose what may prove the biggest 'oil' cost of all – a clear decline in American influence and power almost everywhere in the world. The USA remains by far the world's largest military power, by far the world's biggest economy and of course by far the world's biggest oil consumer.

But the trouble is that bigness no longer equates with power. That is the central consequence and lesson of the information age. Believing both their own propaganda, and that of their enemies and critics – that America is the mighty hegemon, the only hyper-power – the Washington policymakers have fallen into a trap of assuming they possess, and can wield, power they no longer command.

Each attempt to do so worsens the American position and adds to the turmoil in the Middle East, and not only in the Middle East but in other oil-producing areas of Africa and Central Asia and now in South America, which is drifting out of the American orbit.

The point about declining American power and influence, or to put it the other way round, the rise of Asian power and dispersed non-state power, is so central to the world energy scene, now and in the future, that it merits a digression.

There may have been those who, towards the end of 2006, thought that the so-called Iraq Survey Group, under the chairmanship of former US Secretary of State James Baker the Third and former Democrat congressman Lee Hamilton, was going to change the direction of policy in the Middle-East.

But it turned out that the 'new' thinking from this distinguished group of Washington elders was still trapped in the same flawed policy 'box'.

Their united and rooted misconception was, and remains, that America is still in the driving seat in world affairs and still has the power to transform the Middle-East region. Thus, whether one is listening to President Bush himself, or to the Survey Group members, or to all the critical commentators and columnists in Washington, or New York, or indeed in London, there is the same underlying and false assumption – that America may have got things wrong but it is America which must now take the lead in putting them right.

What few of these leaders or politicians or experts have grasped is that size and sheer military weight and spending no longer equate with power and influence in the world. Whatever conclusions the policymakers reach in Washington, whether to 'stay the course' or change direction, will no longer shape events in the Middle East and no Americans, whether in the White House or in Congress or anywhere else, are in a position to control the pattern of events.

Thus there is something almost tragic – and certainly very dangerous – about the persistent belief that the USA can call together a Middle-East conference of nearby powers, such as Iran and Syria, as well as Jordan, Egypt and Turkey, as though it was the victor in some great battle and was now grandly ready to settle the peace terms all round.

The missing piece of understanding in the views coming out of Washington, in the Bush-Blair press conference, in the media, and in the comment from the pundits in newspapers both sides of the Atlantic, is that the age of the microchip has changed everything. It has dispersed power massively – away from the American giant and into the hands of the smallest and most lethal unit, into the most vicious cell and into the most malign clique. America can neither any longer lead the herd nor stand apart from and above the herd, whether the issue is energy security or national security in other senses.

No one puts this better than Thomas Friedman, who in his inspiring book, *The World is Flat*,[19] reminds everyone with marvellous verve that the world has moved from 'a primarily vertical (command and control) value-creation model to an increasingly horizontal (connect and collaborate) creation model'. Apply that not just to business but to power and politics and the world according to Washington vanishes. It is an immensely hard change of mindset to engineer, and even

Friedman, amidst all his insights, falls for the yesterday yearning that there is still somewhere a crash programme to be conjured up which would make America energy-independent. Dream on.

The cross effects on the world energy scene are already devastating. More and more oil-producing regions are becoming too dangerous to invest in. New conflicts on new battlegrounds are opening up in the violent world of oil production, oil pipeline transmission and oil marketing. Even the slaughter in Darfur in western Sudan is partly related to Sudan's emergence as a substantial oil producer.

These ought to be matters where the most intense international cooperation should be developing, both between the big consumers and between consumer and producer nations and regions. The interests of all in both security of supply and security of demand are, after all, identical.[20]

Instead the scramble for oil is creating new tensions, while more and more existing oil production and transmission facilities are becoming vulnerable to terrorist attack. The miniaturization of weaponry in the microchip age has tilted the balance of military strength away from size and towards a patchwork of groups, state-sponsored or non-state, often tiny but now empowered to cause colossal damage.

Money is no object. A prodigious irony is at work. Higher oil prices, driven by geopolitical fears, create rivers of 'petrodollar' cash for the likes of Iran and either other 'rogue' states, or 'rogue' elements within those states. In 2006 it was estimated that an extra £167 billion of oil revenues, on top of an estimated £340 billion the year before, passed into the hands of the main oil-producing states. The cash corrupts but it also buys advanced weapons which in turn arm the terrorist echelons and create still more fear and mayhem, and the oil price lurches further upwards. The Hizbollah movement in Lebanon is a good example of this kind of beneficiary, receiving a generous flow of up-to-date weaponry from oil-rich Iran, via Syria, with which to attack Israel and create internal mayhem within the Lebanese state.

The summit of all fears is the proliferation of nuclear weapons in irrational and irresponsible hands. More irony is at work here. The way out of the oil-dependent age, as well as the way to a low-carbon future, may be in due course eased by a big further expansion all round the world of civil nuclear power. But civil nuclear development also paves the route, via uranium enrichment, to nuclear weapons. This may now be happening in Iran, and may have already happened in North Korea. So the path of peace collides with the path of war.

Conclusion

To be met and handled effectively the clear and present threats to safe and reliable energy systems the world over must be recognized, understood and addressed on a major scale by producers and consumers of fossil fuels – both sides having a direct interest in security of supply, in stability of prices and in a speedy and smooth transition to a new energy pattern.

To recapitulate, the present oil-related dilemmas spring from five sets of events and pressures. The way they are now handled will be of critical importance, not just for current world economic growth and international stability but also for eventual progress in climate control.

These five 'drivers' are: 1) the increasing political instability of oil-producing nations; 2) the probable 'peaking', not of all oil but of the 'old' and cheap oil from the giant Middle-Eastern fields; 3) the huge and widely unforeseen growth in Asian oil demand; 4) the investment failures of the 1980s and 1990s, leading to severe short-term supply constraints; 5) the innate tendency of oil markets to overshoot violently.[21]

Business as usual, or hoping that the wheel will turn and we will be lifted out of danger, as in the past, is not on the cards.[22] Nor should anyone be fooled by sudden drops in prices (whether of oil or gas). The substantial growth in world energy demand, current and potential, against a backdrop of terror and climate concern, differentiates today's energy dilemmas from the oil shocks of the 1970s and 1980s. Markets will drive change, as always. But governments, and the public who elect them, will have to define and set the goals, not to try and outsmart market forces, at which they can always fail, but to ensure that market mechanisms can work properly in a highly politicized environment. Some of the paralysing uncertainties which bedevil almost all short- and medium-term power investments will have to be resolved. A non-catastrophe pathway must be mapped, starting from here, to a more liveable world.

Those in a position to do so must tell the truth, frankly and candidly, about what is already happening, what is shortly to come, and what steps are needed, however painful, to avert much greater disaster for all societies, rich and poor, struggling and advanced, over the next ten years – which is the high danger period. The failed attempts of the 1980s to open a dialogue between producers and consumers, and to recognize that security of supply and security of demand go hand in hand, must be revived and tried again. Oil, which was previously the source and lifeblood of the modern world's strength, is now the source of its weakness and potentially fatal vulnerability.

Later chapters point ways out of the energy labyrinth, however dark the scene now looks. Saving energy, especially oil and oil products, and substituting new energy sources are the two keys to hand in winning through. Ahead, if the world can get there, and given the right leadership and the right framework of global cooperation, lies a potentially golden panorama of new opportunities, new patterns of experience, more comfortable lifestyles, greater climate stability and real advances, instead of paper hopes, in the poorer countries of the developing world.

It is all possible and it is all achievable, without revolution and without relying on distant and utopian goals, however worthy. A way through can certainly be navigated, without worldwide inflation, without major suffering, without a juddering halt to world growth, without conflict. But the need for action, national and international, and on a scale seemingly not yet appreciated, starts now. It is almost, but not quite, too late.

CHAPTER THREE

RETURN WITH
A VENGEANCE

❧ ❧ ❧

There comes a moment. Why steps to ensure Energy
Security must come first and foremost. The tripwire at
our feet. Separating the immediate short-term
dangers from the long-term hopes. Why visions and
warning lectures, carbon trading and other schemes
will not get us through. The *danse macabre* between
energy and geopolitics which cannot go on.

❧ ❧ ❧

We ought to have started long ago. The dangers are not ones
that can be mulled over, reviewed and put in the long-term
planning box. They are immediate and growing daily. The
climbing world market price of crude oil (although it may ease
a fraction from day to day, almost with the latest ups and
downs in world news) tells part of the story but not all of it by
any means. Close behind higher oil prices along come higher
prices for the myriad range of products from oil, from products
dependent on oil and from the other conventional energy
sources which are all interwoven into the energy supply chain.
Along also come higher gas prices, as users switch from oil to

gas and gas demand soars. And along comes huge uncertainty about which new energy technologies to back and which to invest in.

Gas suppliers and industry regulators are fond of pointing out that in the longer term there is plenty of gas available, in the longer term new pipelines will be completed and in the longer term prices may ease,[23] barring, of course, what the gremlins called 'unforeseen events'.

They are right about that. It ought in theory to be impossible to engineer a shortage or scarcity of a resource such as gas which is in such large abundance from so many origins round the world. But somehow the energy planners have managed it. (See Chapter Five for details of the way in which policymakers and Ministers in the UK, one of the most favourably placed nations on earth in the world gas supply system, nevertheless allowed serious shortages to develop in the winter of 2005–06.) It is the short term we live in and it is events far removed from the global gas supply chain which send prices quivering and disrupt daily life and work.

Gas prices have already rocketed in the major advanced countries over the past two years – and then sagged with equal suddenness. Without warning they increased in the UK four times over one winter weekend – from 70p a therm (approximately 100 cubic metres) to £2.40 a therm. Meanwhile, doubled gasoline (petrol) prices have raised all transport costs sharply: air tickets have risen, with surcharge piled on surcharge, electricity bills are much, much higher, and so are household gas bills. All manufacturing processes using either gas or petrochemical products (which is most manufacturing) are facing higher costs. Slight declines in gas prices will bring only temporary relief.

But oil leads the way. Oil enters into everything, because transport costs enter into everything. The civilized and industrialized regions of the world – and now most parts of the developing world as well – are totally, utterly, deeply dependent on oil. Are they also inextricably dependent? We shall see.

The brilliant British columnist and polymath Bernard Levin once wrote a piece explaining how our everyday language, expressions and phrases derive, far more than we realize, from

William Shakespeare. We might be largely unaware of it, but Shakespeare's phrases come into most of everyday speech. In a way it is the same with oil. Our daily lives and habits, our health, our meals, our clothes, our work and pastimes, our homes, our surroundings, our pains and pleasures depend far more on oil and oil products than we care to realize.

Some surprising things made from oil

- Ammonia
- Anaesthetics
- Antihistamines
- Antiseptics
- Artificial limbs
- Aspirin
- Awnings
- Balloons
- Ballpoint pens
- Bandages
- Cameras
- Candles
- Car batteries
- Carpeting
- CDs and cassettes
- Clothing
- Computer chips
- Crayons
- Credit cards
- Deodorants
- Detergents
- Dishwashing liquids
- Disposable nappies
- Dyes
- Electric blankets
- Electric wire coating and tape
- False teeth
- Fertilizers
- Fishing lures
- Fishing rods
- Floor wax
- Food preservatives
- Garden hoses
- Glue
- Golf balls
- Guitar strings
- Hair curlers
- Hand lotion
- Hearing aids
- Heart valves
- Hospital and dental equipment
- House paint
- Ink
- Insecticides
- Insect repellent
- Lavatory seats
- Linoleum flooring
- Lipstick
- Loudspeakers
- Luggage
- Mops
- Motorcycle helmets
- Movie film
- Nail polish
- Numerous components of aircraft and all vehicles
- Oil filters
- Paddling pools
- Paint brushes
- Paint rollers
- Parachutes
- Perfume
- Plastic wood
- Refrigerator linings
- Roofing materials
- Roller skates
- Rubbish bags
- Safety glass
- Sellotape
- Shampoo
- Shaving cream
- Shoe polish
- Shoes
- Shower curtains
- Skis
- Soft contact lenses
- Synthetic rubber products
- Tap washers
- Telephones
- Tennis rackets
- Tents
- Toothpaste
- Tyres
- Umbrellas
- Unbreakable dishes
- Upholstery
- Vitamin capsules
- Water pipes

SOURCE: 'A Thousand Barrels of Oil', Peter Tertzakian, McGraw-Hill, 2006.

When the oil price climbs, some monetary economists like to explain that these extra costs can be 'absorbed', as long as the money supply is controlled (which of course begs a legion of questions about how you do that and how you define the money supply).

That is why, they argue, a doubling of world oil prices can take place without disaster and without a dramatic impact on world economic growth. The demand for oil and its derivatives, they like to say, is 'inelastic' – that is, it does not alter much, or only very slowly, when the price of oil rises. Petrol tanks go on having to be filled, even at £1.20 a litre. Homes have to be lit and heated. Factory processes have to continue. Food has to reach the shops. All that happens – in theory – is that the money has to be found by cutting down on something else. It feels, say economists, ever eager to simplify, just like paying higher VAT or sales tax on a product, although the revenues, or most of them, disappear into petrodollars rather than into national coffers, where at least in theory some say over them is retained.

But what if everything else is already cut down? What about the poorest, and indeed all families on tight budgets (which is most people), who have no room whatever to cut any further, or to fork out for this painful extra 'tax'? Well, say the experts, sounding ominously like Marie Antoinette, the old must wear two woolly pullovers and if necessary there must be higher welfare aid to meet more fuel poverty, and therefore higher taxes to finance bigger government budgets to pay for it all.

That may be the chilly economic logic of higher fuel and energy prices. The reality is less accommodating. Admittedly there are other factors also at work to 'soften' the impact of higher oil prices in some markets. In Europe the already existing heavy fuel tax on petrol masks the basic rise in the price of the delivered petrol or diesel product at the forecourt pump.[24] Another factor is that oil is still priced in dollars and when the dollar is weak, as in recent years, this further offsets the cost to non-dollar consumers.

But there comes a moment, a tipping point in the current fashionable and rather good phrase, and it could be about to come now. Nobody can be sure because we are talking about those 'unforeseen events' which invariably occur and to which the world's energy supply chains have been allowed to become totally vulnerable and exposed. The point at which consumers throw over old lifestyles and preferences, or turn on leaders who have let the situation drift and given no guidance about the immediate way out of the labyrinth, may not be immediately obvious and it may commence in a sort of patchy, dribbling way. Or it may come with a sudden collective rush, a universal realization which could easily trip over into panic.

Because the energy issue is so all-pervasive, and because it is tangled up with so many other key developments, whatever happens the transition will look confused and bewildering. Long-term and short-term trends will seem to be jumbled together. The subject of energy has become deeply interwoven not only with international security but with all the hopes and fears about global warming and violent climate changes which already seem to be happening. As noted, many leading political figures, when asked about energy security, immediately launch into warnings about the greenhouse effect and the dangers to the planet this will present unless we can cut carbon emissions and stop burning fossil fuels.

Yet the threats of serious disruption to energy supplies and oil-dependent living are immediate. This is because an agonisingly tight current supply/demand balance is already with us, because world oil demand continues to soar, because infrastructure investment decisions have to be made now to ensure power supplies and because this all coincides with, and interacts with, a deeply disturbed world full of new dangers which a few decades back, in the politically and ideologically frozen days of the Cold War, looked remote and unimportant but now affect oil supplies by the hour. Green hopes beckon down the road but geopolitics are hammering at the door.

Global warming is already having a dire impact, with increasingly violent weather patterns. If, as seems quite likely, this is the consequence of man-made activities and human disruption then the damage for the next three or four decades is already done, due to the amount of CO_2 which our forbears pumped into the atmosphere. The accumulation of CO_2 for a period from circa 1840 to the present day, most of it emitted by the USA, Russia and Europe, ensures that all present and future efforts to curb carbon, while commendable and necessary, will only have an impact after, say, 2040 at the earliest. Sir David King, the UK Government's Chief Scientific Adviser, is quite explicit on this. In his Greenpeace Business Lecture of October 2004 he explained that 'because of the current levels of carbon dioxide in the atmosphere, the climate change effects are going to be with us for the next 30 or 40 years whatever we do. Even if we stopped emitting now we have effectively committed ourselves to a further period of global warming'.

Geologists are fond of saying that a million years is a short time. Human beings have only been around for a tiny blip at the end of the 4,500 million years since the planet first formed. The time frame of existence of the human species may only be a few hundred thousand years. Most mammal species have faded or vanished after that sort of span.

As tenants of the planet we are still capable, by our behaviour, of wrecking the home, of blowing up the environment, disrupting its cycles and rhythms so violently that it spews humanity out prematurely. We are still all too capable of desecrating the loving, living planet – Gaia, as James Lovelock so tenderly portrays it. A long series of books by Lovelock, from his 1979 work *Gaia: A new Look at Life on earth*, through to *The Revenge of Gaia* in 2006, have had a profound influence on the way the planet's future is seen. Lovelock now believes that 'clean' nuclear power could save the day. But will governments and societies wracked by energy insecurity ever muster the confidence again to commit themselves to these giant edifices which take decades to construct and arouse so many public fears?

The Emaciating Drug

All this is undeniably central and vital, even if there are big uncertainties and scientific disagreements about the detail. It is good that present generations should act now in a way that past generations did not, so that our grandchildren, great-grandchildren and their progeny have a better chance of living in a balanced and harmonious planet (human nature and other things allowing).

But energy security must be the first and immediate priority. Well-meaning energy planners and political idealists who keep putting all fossil fuel burning on the blacklist are endangering both short-term and long-term goals. Government committees meet and deliberate on sustainable development. But how much will be left to sustain? A world still gripped by the oil 'addiction', and emaciated by the oil drug, will be far too weak to tackle the challenges of global warming, or accept the extra burdens and costs on behalf of their grandchildren. The happier days will never come. Instead they could slip further and further out of reach.

That, too, is why the old cry of many oilmen for more oil to be produced, still echoed by some political leaders, is such a devastatingly misguided call. When the American President calls for more indigenous oil production, and for faster exploitation of the Alaskan wildlife regions, or when the British Chancellor beseeches OPEC members to increase both output and refining capacity,[25] they are simply calling for more of the drug, to reinforce the addiction, and diverting attention from the real and near-term dangers. They are also calling for the impossible, since most OPEC members are already extracting oil from their giant fields at maximum rates consistent with good oilfield practice, where they are not anyway held back by political turbulence and sabotage threats.

In short, neither America nor the world can drill their way out of danger. Neither the myopic, backward-looking 'leaders' who hold these views, nor those longer-term visionaries and idealists, dreaming of a fossil-fuel-free world, are facing up to the current realities or fulfilling their responsibilities in preparing our societies for the challenges lying directly ahead.

Green Dreams – and Movies

New thinking and new attitudes are certainly required urgently. But the green visions must not be allowed to get in the way of reality or divert minds from hard, immediate challenges.

The green appeal is very strong – a sort of beautifully photographed Hollywood cameo production, suitable for all ages. It can be played upon many tunes. Al Gore's already-mentioned movie takes one approach, and puts him firmly in the apocalypse camp (where a crowd of authors are already gathered).[26] But here's another:

> As the background music rises, our hero and heroine scramble out into the calming daylight of reliable, and reasonably cheap, energy supplies, produced without wars and political storms, without pollution or frightening damage to the balance of the great green planet, without blackened landscapes, and walk hand in hand towards the rosy-fingered dawn of a new and better day. Our beautiful planet, a living system of dazzling intricacy, is saved.
>
> Behind them our two leave a collapsing world of global desecration, pollution, oil rivalries, vast and vulnerable energy supply systems, soaring prices, power cuts and black-outs, petrol queues, foreign policy conflicts distorted by fear and by the desperation of governments, corruption and yawning divides between super-rich and poor (gaps in which extremism and fanatical violence breed).
>
> All this and more cracks and crumbles into the sea, like the great palace of King Minos, swept to destruction by the largest volcanic eruption of all time,[27] while in the background giant explosions light the sky, finally destroying the idiocies and errors of the oil age and swallowing up its relics.
>
> Our two characters turn and look back at the blazing citadel, their faces blackened, their clothes torn, but safe. Then they embrace and go forward. The music reaches its crescendo. The curtains close. The movie is over.

That is the big screen version. Outside the cinema things may be a bit different. Transport the hero and heroine back into the real world and let's suppose the hero is a slightly older man, greying nicely at the temples, while his girl companion is a young idealist, full of determination to build a better world.

Out in the daylight they clearly have some differences. The younger one, the girl, probably believes, quite sincerely, that there really could be some sort of happy ending if only the older generation had not made so many stupid mistakes and missed so many opportunities over the past 30 years, and if they would stop making still more mistakes now. She has read the magazine *Greenpeace*, listened to Sir Jonathan Porritt,[28] seen the Al Gore movie, absorbed at least the conclusions of Sir Nicholas Stern's massive economic analysis of climate change, and heard all the political leaders, both right and left, on the dangers to the planet of fossil fuel burning.

The older man thinks his lot have done pretty well in coping with the oil shocks of the 1970s, 1980s and 1990s. He still believes that market economics and business enterprise can see the world through another oil crisis, that the policies of the later part of the last century were appropriate to the time and effective and that most, if not all, government intervention will probably make things worse. Oil scarcity, like most scarcities, is usually succeeded by oil glut, as happened last time. It could happen again.

But he admits that his companion has a point. Things have changed radically. Markets still work, but energy supply and oil supply in particular are encased in the deepest and most disruptive politics. The world oil market in particular seems hair-raisingly fragile and volatile. There is a totally new energy environment and this is plainly infecting the whole international scene.

The girl observes – a bit sharply – that the energy policies of past decades have led to catastrophic results, that we are now paying a terrible price for past errors and lost opportunities, and could pay a bigger one still unless we come to our senses.

Energy security – that is warm homes, lit streets, humming factories – is under threat as never before. Oil prices are sky high, even if easing from their peak from time to time, apparently dragging up gas prices with them. Dependence on oil is greater than ever, and dependence on the unsettled Middle East greater still. There is a lot of talk about a low carbon future. But the green 'alternative' strategies still seem very elusive and futuristic – except for wind energy, where the green contribution may be growing modestly but the price is being paid in terms of forests of giant wind pylons desecrating remote landscapes, carrying heavy subsidies which are making some people very rich, and killing beautiful birds (as well as being unreliable as base-load electricity supply sources). Most alternatives still look more expensive than a barrel of oil, even at $60, and of course much more expensive still if and when oil sinks back to $30 to $40 a barrel. And as for carbon gas emissions, these are still rising fast, targets are being missed and the ice cap is melting, stranding baby walruses and threatening polar bears.

Both are right and both are wrong. The alternative green scenario probably would work for most residential consumers in the fairly distant future, provided that industrial needs were met by relatively carbon-free nuclear power, and providing the whole world, developed and developing, joined in. (The younger one is not so sure about this. It's a bit of a dilemma.) And the older man's conjecture that eventually oil prices may well ease, if not actually collapse as they did before, is probably right.

But they are both wrong about today, tomorrow and probably at least the next five years. Why? *Because energy has become hopelessly entangled with politics, with extreme Islam, with fanaticism and terror and with the global redistribution of power.*

In the long run, say the idealists, it will all be OK. Well, yes, but not on present trends and not with present priorities. Perhaps the Pied Piper gave the same assurances as his strangely seductive tune rang out – and remember Keynes again. It is in the next five to ten years that things are going to turn really nasty. Policymakers seem to find these prospects too awkward to talk about. Far easier to discuss dreamland targets 40 years ahead.[29]

This explains why official documents trying to set out the energy scene are full of astounding and quite unresolved contradictions. For instance, the UK Energy Review and Report, *The Energy Challenge* (already referred to earlier), gives top slot priority to the carbon challenge but then calls for big new plans for gas and coal investment. It calls into being a world of cleaner energy and reduced carbon emissions and then admits that 'fossil fuels will constitute the majority of our energy mix for the foreseeable future'. Carbon must be reduced but carbon emissions are rising. Oil dependence must be curbed but – and here defeatism takes over completely – 'it will take decades before we see a real shift away from oil as the predominant oil source'. And the answer? Why, of course, 'more demanding targets' pitched further and further into the blue yonder.

Now pause while we examine some of the realities which tend to go unmentioned or skimmed over when the policymakers and their advisers speak, or when the film-makers and publicists take up the subject.[30] Below are seven awkwardnesses, or 'inconvenient truths' (Al Gore again), which somehow do not feature much in official pronouncements.

A Few More Inconvenient Truths

ONE. Far from the oil age being over, the world is more dependent than ever on oil and according to official estimates is set to become more dependent still on the shaky Middle East.
It would of course be nice if it were otherwise. The 'post-oil age' makes a good subject for TV and radio series, and for cascades of books about the end of oil.[31] Most of these writers are strongly influenced both by the history of energy, for example the way in which the energy 'throne' passed from wood to coal, and from whale oil to rock oil, and now presumably to 'something else', and by the already noted theory from M. King Hubbert, the respected Shell geologist, that world oil production is peaking any time now.

But it is probably a mistake to think in terms of 'ages' any more at all. The eventual future is going to have a place for all these energy sources – except oil from the poor sperm whales – with coal making a massive comeback and even the right sort of wood playing a new part, as Sweden's plans demonstrate.[32] The future is not going to look like, or evolve like, the past at all. The future, as Alvin Toffler keeps reminding us, is non-linear, and that applies to energy as much as elsewhere. What the ancients called Fate will always intervene.

Besides, the Hubbert peak concept – that oil reserves are running down, and eventually out – is fraught with definition difficulties. It all depends how reserves of oil are categorized and whether actual deposits of economically accessible oil, at current or likely future prices and with current or likely recovery technology, are replacing what is being extracted, or not. And as no one knows where the politically driven price will go next, and very few people are sure about the curve of new technology, the whole concept is wild surmise, probably creating more dispute than clarity.

The weasel word in all this opining is 'eventually'. World demand for oil and all its products is currently growing faster than ever – far above many predictions[33] – and unless diverted will continue to do so for at least the decade ahead. Daily world oil consumption has now reached 86 million barrels (2006), or 1,000 barrels a second. Consumption has never been so high in history. The latest forecasts from the International Energy Agency (IEA) in Paris are for a global oil thirst by 2015 of 99 million barrels a day, and 116 million barrels by 2030. The oil is there in the ground. Whether politics, economics and technology allow the necessary production infrastructure and capacity to be built and operated is quite another matter. Only three years ago (in 2004) the IEA was forecasting 81 million barrels a day by now (in 2006–7). Back in the 1990s analysts were predicting levels of 65 million barrels a day by 2005 – all hopelessly wide of the mark. Could they be wrong again? Yes they could.

Production to meet even these colossal volumes of demand could only be reached by still more fields being opened up in the present key producing areas in the Persian Gulf and from new and much more remote, expensive and dangerous regions. Whether this can be achieved or not, the net effect will be that world dependence on the unstable Middle East will increase (some say double) in the next ten years. That is why there has to be a change of direction in energy policies. The predictions are pointing to the impossible. To borrow a phrase from the popular British actor Michael Caine, not a lot of people know that.

TWO. There is a head-on conflict between the growing energy needs of the poorest half of the world and the worries in the already industrialized countries about climate change.

Much the biggest daily imbiber remains the USA, despite having achieved considerable reductions in oil consumption in the past. The big American shifts downwards of the 1980s between GDP growth and oil consumption growth are still having some effect. If the Americans had continued on their pre-1980 path they would now be drinking 30 million barrels a day instead of 20. But meanwhile it is the thirsty Asians who are the new and demanding guests at the oil feast.

For example, the Chinese, whom everyone (including China's leaders themselves) thought would by 2006 be self-sufficient in oil, are importing four million barrels a day. The previous year it was three million, in 2007 it looks like five. The EIA thinks it will be ten by 2012, but they could easily be way out on that, too.

The Chinese worry a lot about this, but from their perspective things look different. The immediate polluting effects of industrialization are obvious enough, with yellow palls of smoke hanging over major Chinese cities. Getting rid of these is an immediate incentive to clean up coal-fired power operations and reduce sulphur emissions.

But when it comes to climate change the understandable reaction, not just from China but from the whole emerging world, is that while abrupt climate shifts could damage everybody, the primary responsibility lies with the already-industrialized nations whose past behaviour caused it all. They are the ones who should be paying now for a cleaner future.

The situation has been well summed up by the Indian Minister of the Environment, A.J. Raja. At the official opening of the Asia-Pacific Partnership on Clean Development and Climate Change in 2006, he said: 'We are developing countries, we have our own agenda for our development activities, so we cannot give any promise, any commitment, to reduce further our emissions'. Just so.

For the richer nations to start lecturing the would-be developing ones about energy consumption, or even proposing globally organized penalties for excessive carbon emissions in the poorer world struggling to catch up, sounds from this viewpoint thoroughly unreasonable and unfair. Vulnerability to oil shortage is one thing, and the Chinese dislike their growing dependence on uncertain oil as much as anyone. But energy there must be to power China's rapid growth.

Massive steps are therefore being taken to reduce future oil import vulnerability, not least building some 560 new coal-fired power stations fast (one new one roughly every ten days between now and 2012) – hopefully, but by no means definitely, using clean coal technologies – and no less than 40 new nuclear power stations are planned. China is also building 30 coal liquefaction plants (see Chapter Three) and the vast Three Gorges hydroelectric project will be completed by 2014, although whether this is any more of a plus or a minus in environmental terms is open for debate.

But in China car fuel (petrol or diesel) is hardly taxed at all. This would at least be understandable, although wrong-headed, if car ownership was at world average levels (120 vehicles per thousand people). But it is 10 per thousand today in China. This is the time to act, while it is still politically possible for a regime which is not all that secure and while the damage to China's amazing dynamism would be minimal. By the time car ownership in China gets even near to half world levels, at say 50 per thousand (which will still only be one tenth of the current EU ownership level), that will put 150 million vehicles on China's roads, instead of the present 30 million, multiplying China's oil needs by dizzying amounts and of course paralysing any political will on the part of a nervous ruling establishment to raise petrol taxes and offend such a vast constituency.

Already Chinese cities are experiencing blackouts, brownouts and gasoline shortages almost every day. About 300 million Chinese are now beginning to enjoy a sort of middle-class wealth level. Waiting in the wings are another 800 million of whom at least half want to head towards city life, towards car ownership, electric power, gas for cooking and heating and other comforts. Even if everything else was calm and peaceful in the energy world, and even if a green future of renewable and sustainable energy was just around the corner, these trends alone would guarantee a ferocious and unending upward pressure on global resources – and not just of oil and gas but of every kind of mineral and raw material as well.

Aware of this prospect the Chinese state-owned oil companies have been scouring the globe in an attempt to tie up future secure oil supplies. In Angola CNOOC paid ten times the nearest private sector company offering for offshore oil drilling concessions. In Sudan elaborate long-term contracts have been struck.[34] In Venezuela President Hugo Chavez, always eager to snub the Americans, has signed long-term supply contracts with Beijing. Nigeria has done the same. Saudi Arabia is now China's largest overseas oil supplier, providing about 15 per cent of China's imports, with Oman, Angola, Iran, Russia, Vietnam and Yemen providing another 60 per cent.

None of this will protect the Chinese oil consumer from violent price rises when world markets receive their next shock. This the Chinese authorities must realize, but meanwhile they remain under intense pressure to try all avenues to meet their nation's soaring energy thirst.

That is China. Add in India, where the population will exceed China's shortly and where oil consumption trends are following along behind at a similar rate. Add in the rest of the developing world and the IEA estimates for oil consumption by 2020, currently about 120 million barrels a day, begin to look unrealistically modest.

Against the background of ever-rising oil thirst the supply and demand outlook would be bound to remain very tight, even in a peaceful world, not because of lack of resources in the ground – there are plenty of those left – but because the giant fields in the cheap and easy 'desert' regions are running down and also getting much more dangerous to invest in, while the more remote unexplored fields are getting more and more costly to operate. More and more water is coming out of the ground with the oil in the 'cheap' desert fields, a sure sign that these wells are running down. So it is these 'easy' reserves which are not being replaced, or being only partly replaced, and then by oil from more difficult areas which costs many times more to extract.

Higher-priced oil ought to work the other way. It ought in theory to lead to lower oil prices in the next stage of the cycle

by curbing consumption and getting the supply-demand equa-
tion back into balance. But consumers across Asia, and even
more in the oil-producing countries of the Middle East themselves, are largely shielded from higher prices by minimal
taxation on fuel, or even by subsidies.[35] The overall offsetting
effect on global oil demand is going to work its way through
very slowly, and may anyway be overwhelmed by insatiable
Asian oil thirst. As the economists and econometricians like to
say, another variable has been introduced to the supply-
demand mechanism. The cycle pattern has changed. Someone
has bent the cycle's wheels.

The hardest-hit victims of high oil prices are those who are
already on the edge of survival and subsistence. The resources
needed to develop and manage water resources, to raise health-
care standards, to open schools and build roads, are draining
away fast into petrodollars. Some of these are being recycled
back into development aid, but most are not, and there is no
guarantee anyway that aid funds will end up in genuine development or improvement. The more likely destination, all too
often, is Swiss bank accounts.

The dilemma raised by energy and development is intense,
and the policymakers are stuck firmly on its horns. Any hope
for raising the living standards of billions depends on a massive
and rapid increase in energy use. The cheapest energy is the
kind that is going to be used. The present pathway leads
straight via oil and gas (still cheaper than renewables), and via
coal, the cheapest of all, to much larger emissions of CO_2 and
still more climate dangers in the distant future. Together,
China, India and America contain half the world's coal reserves
– enough to last centuries ahead. They intend to burn it to provide the vital energy needed for their economic advance.
Somehow that pathway has to be changed or modified without
penalizing the poorest, which means a complete break with
past aid and development strategies. Not many people seem to
know that either – although one of the few who does is
Hernando de Soto, whose understanding of the crucial role in
the development process of clear property rights and energy
supplies to the home through efficiently organized utilities has
been largely ignored by the aid lobbies.[36]

THREE. Worldwide carbon emissions are still rising very fast. We are not doing for future generations what we say we should be doing. (Even if the UK, which accounts for less than 2 per cent of worldwide emissions, was getting some real results, it would make zero impact on climate change.)

Authorities round the world aver that carbon omissions are going to be reduced and global warming consequently slowed down. Maybe. But that will be decades ahead. Carbon emissions are still climbing fast and all the nearer-term targets for limiting emissions – both those set by the Kyoto Protocols and the tougher ones adopted by the European Union – are going to be missed.[37] The UK 2006 Energy Review, already referred to, ruefully acknowledges that, despite all the efforts and hopes to curb carbon emissions, they are still rising in Britain. The International Energy Agency, with brutal frankness, opines that world carbon emissions will be up 55 per cent by 2030.[41]

Even if the big carbon-emitting countries pull their weight (and as has been shown, they have precious little incentive to do so),[39] the clear evidence for the next 30 to 40 years is that there is nothing, but nothing, to be done to prevent the heating up of the ocean's surface by the CO_2 already present in the atmosphere and the consequent violent changes in the world's weather patterns. The grim situation is pre-programmed. That is why voices urging much more vigorous adjustment and adaptation to climate change, as well as efforts to mitigate, are so relevant – and so unpopular amongst carbon-cutting crusaders who cannot bear to hear this awkward reality.[40]

Numerous highly elaborate schemes have been introduced to price carbon, and to cut future carbon emissions (the carbon 'flow' rather than the already existing carbon 'stock'), such as the EU emissions trading scheme. Unfortunately, the price of permits to emit carbon keeps changing. Some countries have given away too many, some have given too many exemptions. Some are proposing to auction carbon permits instead of just issuing them. All this leads to paralysing dilemmas for businesses planning how much to invest in cleaner energy sources. And the incentive to move carbon-generating processes to countries with easier emission controls, or none, keeps growing. The net effect on the amount of carbon entering the

A giant carrier of liquid natural gas, compressed and frozen. The fuel of the future? Or a terrorist's dream target? And will low-carbon campaigners welcome it?

atmosphere is therefore miniscule. The EU Emissions Trading Scheme has been, in the words of *The Economist,* 'in serious trouble,' with absurdly loose targets and lavish printing of permits allowing carbon emissions to rise. The goody-goody UK has the tightest controls, which means that its industries have been paying a fortune to other member states to buy their permits (and making some people in these countries suddenly and delightfully rich). Some repair work may have now improved the situation and tightened the permit issuing process Europe-wide. But the inherent weaknesses in schemes of this kind remain.

A far more effective approach to carbon reduction for the coming generations would be a carbon tax which would hit directly at the worst polluters wherever they were located in the planet – and that probably means increasingly in China and India. Another method, which could help on a small scale, is carbon offsetting – a grass-roots scheme which requires *all* emitters of carbon, including individuals, somehow to calculate the effects of what they are doing and make compensating payments which are then, in theory, used for carbon-saving investment in developing countries, or in trees being planted somewhere. It may or may not help the next generation, but here too people will want to know very clearly where their money is going and how it is being spent. They may have to wait a long time and they may be doing no good at all.

The lack of enthusiasm with which any global plans for carbon pricing are likely to be greeted in industrializing countries struggling up the ladder has been noted. But governments everywhere have the task of 'selling' the right measures and using plausible and soundly based arguments in doing so.

Carbon penalties, or taxes, are yet more additions to the already high price to the poor consumer of conventional fuels, or products, including electricity, which depend on them.

If people are to be asked to pay more, whether in taxes or levies, they need to be told what they are paying for and where the money will go. Paying to alter the climate in 40 years time (with a good deal of uncertainty about the results) does not sound very enticing. Dressing up the proposition to pretend that carbon curbs and penalties can somehow calm *current* climate changes could have an even more negative reception, when it is shown to be untrue.[41] The Green prospectus needs at all times to be advanced with care and precision. As they like to say, terms and conditions apply.

Realistic and undeluded policymakers should be selling a much more honest product – namely that reducing oil dependence here and now is a practical goal worth paying for, and if this has to be done by making conventional oil-related energy more expensive, and/or subsidizing infant renewable energy sources in its place, these are results which can and should be delivered pronto and are value for money. Some intervention to set the right context and goals in which market forces and enterprise can operate to achieve this end is justified. That this approach would also eventually deliver big carbon savings is an added benefit.

Meanwhile, bodies such as the Carbon Trust in the UK issue plentiful admonitory literature and pay their officers and staffs generous salary uplift bonuses for success. But if their mission is to reduce carbon emissions, or even slow their growth, they are failing, not succeeding. In the UK carbon emissions have actually been rising every year since 2002.[42]

If the low-carbon crusade is pursued almost to the exclusion of more immediate energy dangers, the risk is that it will cloud the issue and engender huge public disappointment and rejection, both in the developed world and in the developing countries. That is why the message must be made many times more compelling and more immediate. Energy security, climate security and the escape from poverty all march together and will succeed or fail together.

Somehow the policymakers and politicians do not find it very comfortable to talk about these things.

FOUR. The oil balance between supply and demand is so precarious that a new shock, which could happen any minute, could send the price leaping further skywards. The world oil supply chain is not getting safer. Today it is more vulnerable than ever before.
In contrast to the position in the 1980s, when there was plenty of available spare oil production capacity, if the Saudis and others cared to use it, today there is almost none. The authorities talk about raising oil production. They talk less about the miniscule margin of spare capacity available to meet the next shocks.

Strategic reserves of oil have been built up in a number of countries, which are supposed to be of some comfort (the US strategic reserve would last for about three months). But the very act of using them sends nerves jangling and motorists to the pumps to keep a full tank, just in case. And once they are gone they cannot be replenished until the crisis in question has abated.

Meanwhile, President Bush calls for higher oil production inside the USA, as though this will somehow insulate American consumers against world oil prices. But like some Chinese leaders, he fails to understand that oil is 'fungible'. That is to say it is a globally marketed commodity, so that when supply or demand shocks occur the consequences are shared by all.

As noted, the British Chancellor, and would-be Prime Minister, Gordon Brown, has called for OPEC oil members to produce more oil and he is clearly under the impression that this is practicable. In fact it is not. The current supply-demand balance for oil is intensely tight and very precarious. Prices may sag from time to time but any new shock (and plenty are coming) will send prices soaring much further. Efforts in the advanced and more northerly states to meet surging fuel poverty and cold weather hardship will continue to lag far behind events.

Even if world oil demand was to be held at existing levels (which will not happen) huge new oil reserves would need to be opened up to match it, along with sharply stepped-up investment in exploration, development drilling and finally production, all of which would take several years to deliver results. Some of that is now happening but it is all very late in the day. It was back in the 1980s and 1990s that the investment was needed but did not take place.

Today the world still remains totally dependent on oil. Oil enters into all stages of production and consumption, daily living, social activity and society's operation. It is the noose round the world's neck. As the noose tightens violent 'spikes' in prices of both oil and gas (and therefore in petrol, heating oil and kerosene, and all industrial processes) are occurring, and will do so with increasing frequency in the months and years just ahead. These will cause immediate hardship and major economic difficulties for which little warning has been given and few preparations made. Quite aside from possible interruptions to oil supplies the very fact of price volatility is immensely damaging to key industries, such as the car and truck manufacturers (one tenth of employment in the USA), the airline industry and every business involved in, or dependent on, road transport.

The potential sources of shocks to the system are not only multiplying but greatly amplified by the integrated nature of the world's oil supply chain. Thus it needs only an oil spill, or a report of pipeline corrosion, or a hurricane danger, or a temporary oil platform shutdown somewhere – anywhere where oil is being produced – to send oil prices spiralling. BP's decision in mid-2006 to close part of its Prudhoe Alaskan field after reports of serious pipe corrosion added two dollars to the crude price overnight.

But that is just the start. The world oil supply chain is now confronted by new threats coming at it from all angles, over and above these more technical and operational ones. In 19 of the world's leading oil producing states political upheaval is more than a possibility, it is a probability. Each new political disturbance or each new intrusion by greedy governments into the oil-producing sector (such as Venezuela's or Bolivia's nationalization of Exxon's and Chevron's local asset) sends oil prices reeling.

That would be dangerous enough if it was not for the even more serious, and almost ubiquitous, threat of terrorist attack and sabotage of oil facilities. There is nothing remote about this threat – it is being exercised daily. In Iraq all hopes of raising oil production to the levels which a peaceful Iraq could easily sustain (around five million barrels a day) have been prevented by repeated pipeline and facility sabotage, and output is stuck at around two million. In Nigeria pipeline attacks have cut output by at least half a million barrels a day. Kidnapping of foreign oil workers – and in one case, murder – have hardly helped encourage outside investment.

In Saudi Arabia key oil-handling terminals and refineries like Ab-Qaiq have been declared prime targets by terrorist groups. Al-Qaeda has made no secret of its determination to strike at key oil facilities. Oil, says one of Al Qaeda's statements, is the provision line and the feeding to the artery of the crusader nations. Another Al-Qaeda pronouncement (quoted in Chapter One) refers to oil supply lines as 'the umbilical cord and lifeline' to the West. Whether it is an artery or cord or lifeline the message is quite plain, plainer even than the warnings from Hitler in his writings that he was going to attack Germany's neighbours and destroy the Jews (which most of the world ignored at the time). Cutting the artery, as terror groups large and small can well see, is a truly lethal blow in the campaign to eject 'the crusaders' once again[43] from the Middle East and from sacred Islamic soil. What was done in 1215, with the ending of the Crusader Kingdom of Jerusalem and the ejection by Saladin of the infidels from the territory of Islam at last, after 125 years, can and must be repeated, but by different means.

The attack at Ab-Qaiq in the Spring of 2006 appeared to have been frustrated although the attackers got well through the plant's outer defences. Had they planted bombs closer in some, 15 per cent of the world's daily oil exports could have been severed. An assault on Ras Tannura, the world's largest offshore oil terminal, or on the Juaima or Jubail oil complexes, would be even more devastating. These are all heavily guarded, but the vulnerable points are the pipelines which criss-cross the country over huge distances and just cannot all be protected.

The most vulnerable point, as already noted, is the Straits of Hormuz at the mouth of the Arabian Gulf; 18 per cent of the world's oil exports pass through this narrow channel every day. Mining it might be beyond the capacities of a small terrorist group – although the technical means of doing so are getting more accessible all the time. But it is not beyond the capacities of an angry and provoked Iran, whose leaders have already hinted that they might be driven to this expedient if attacked, bringing down the temple on their own heads but sending the world reeling at the same time.

Today's global oil supply chain is more vulnerable than ever before in its history. More shocks are a certainty. There were about 400 incidents affecting oil supply in 2006. Many more are round the corner. Yet political leaders remain curiously reluctant to explain the full dangers or to prepare the public for what is inevitable. Schemes to limit carbon emissions seem easier and quicker to engage the public imagination.

Instead, each new shock and each upward lurch in energy prices appear to catch everyone by surprise and leave policy, and protective and remedial action, floundering far behind. It cannot go on.

FIVE. The measures and lifestyle changes urgently needed will be hurtful and highly regressive in their impact unless new policies are devised very soon in the main consuming countries. The hidden costs of oil dependency tend to stay hidden.
The changeover to a new energy balance may make some people a lot of money, but for the poorest not much will trickle down and the impact, if not prepared for now, will be extremely painful and disruptive. That, too, is not much discussed in public. It will require large supplies of courage from governments and politicians. There are NO instant solutions

or magic bullet formulae. Idealistic lectures about a low-carbon future and the wonders of carbon-free nuclear power distract (and maybe are intended to distract) from the real and immediate energy issues which must be tackled and are not being tackled.

Political leaders do not find it easy to spell out some of the changes that will have to be faced. For instance, since transport accounts for half the industrialized nations' oil consumption (more in the case of the utterly automobile-dependent USA) the entire world's automotive industries have to be overhauled here and now. Super-fuel-efficient, very high miles per gallon vehicles, both trucks and cars, can be made of ultralight materials with existing technologies. Re-tooling should be under way in all automotive plants everywhere, from Detroit to Nagoya, and from Stuttgart to Solihull, from Moscow to Seoul. Engines can be adjusted now to take the vegetable and waste-based oils (ethanol, biodiesel – more in Chapter Six). The move beyond mineral ground oil should be highly profitable, and not costly, but for this to be so the tax system has to be immediately reshaped to remove perverse incentives and shift customer choice swiftly to low-energy vehicles. The incentive to buy efficient vehicles cannot rely on high petrol taxes alone.

Many more details of what is both possible and immediately necessary at the public policy level have been elaborated on by Professor Amory Lovins in the USA in his ground-breaking presentation.[44]

Lovins has brilliantly set out the full and coherent programme of immediate possibilities (more of these in Chapters Five and Six). His themes are set in an American context – and it is in the USA that the big changes must come. But the ideas and detailed proposals for implementation are just as applicable in the UK, and indeed in all oil-consuming societies.

The USA has 4.6 per cent of the world's population, produces 21 per cent of world GNP and drinks up 26 per cent of the world's oil. Its own share of oil production is 8 per cent and falling fast.[45] The smaller nations can prod, demonstrate, lead by example, using the power and influence of the network age. But America must make the changes – in its industries, in its habits, in its wants and in its needs.

Government and the public sector can also act by example, as well as by practical measures which enhance national energy security directly. Governments purchase fleets of new vehicles, every single one of which should have a low consumption engine. In many countries, too, especially the USA and the UK, the military is a huge and lax consumer of oil and oil products. Military establishments tend to be slow to reform and adapt, and even slower to take the lead in innovation and new techniques for low-energy consumption. Had a fraction of the ingenuity and technology going into new weaponry over the last decade been diverted into reducing oil consumption by the armed forces, the budgetary gains, and the gains in efficient military performance and delivery, could have been enormous.

This applies as much in the UK as in the USA. Today the British army, said to be under near-impossible budgetary pressure, still expends vast resources maintaining fuel supply lines to unbelievably thirsty tanks and trucks (more of this in Chapter Six). The potential both for manpower and resource savings in fuel transportation, and in terms of higher mobility and efficiency for front line forces unhampered by the need for such large volumes of gasoline, is very considerable – and largely unexploited.

A curious political silence also hangs over discussion of the large hidden costs of oil dependency and of maintaining the oil supply chain. To sustain the present world energy supply pattern, as at present organized, involves both America and Europe in elaborate and ultra-expensive foreign policy commitments. A gigantic military infrastructure is needed to underpin these commitments, and it is now proving almost unaffordable. This comes on top of the straightforward resource transfer cost as billions of dollars flow into the coffers of the oil-producing states and purchasing power is sucked out of the American economy, still one of the world economy's main drivers.

This process in turn, swelling the petrodollar pool, has long since proved to have highly debilitating effects on the recipient countries, paralysing diversity in development, ensuring lopsided economic growth, entrenching corruption and reinforcing the underground streams of alienation and rebellion which threaten stability even further than at present. This is the curse of oil at work, negating development aid and investment and ensuring a fertile seedbed for youth disillusion and eager adherence to terrorist revolutionary creeds.

None of these crippling 'costs' of oil tend to get openly discussed, or directly connected in public debate with oil dependency, nor are they reflected fully in the world market price for the commodity. Yet they are there and they have to be paid for by the advanced countries, often in blood as well as in money.

The policymakers need to remind themselves, as well as the public, that a galaxy of highly effective energy-saving actions can be adopted here and now, overnight, at city, county, town and village level. This can take place across the entire advanced world. The changes both needed and possible in energy 'habits' interrelate and interact with the transformations which are anyway now seen as increasingly desirable in other areas of life, such as the right relationship between central and local government.

Local initiatives and enlightened self-interest are just as much the drivers in coping with energy security challenges. But first the process of 'enlightening', that is of accurately informing and warning frankly and openly what lies ahead, must be put into a much higher gear. That's not happening.

SIX. Governments are very nearly paralysed about what big, long-term infrastructure investments to back – though they cover it up with talk about options and reviews.

The Minister sits at his desk in the Department of Long-Term Co-ordination and Strategic Planning.[46] He's been told, in no uncertain terms, by both his colleagues and the press, to get on with a plan for reducing oil dependence, cutting CO_2 emissions and ensuring cheap, reliable future energy supplies for all.

He has to approve which alternatives to oil the Government is going to back, either by just voicing general approval, by tax breaks and incentives, partnership with the private sector or by outright funding.

His paralysis comes from the awful fact that any decisions involving investment in energy infrastructure, of whatever kind, lock the economy in for years ahead. Once the fateful decision is made to 'go nuclear, go for coal, go for gas, go for renewables', the die is cast. Growing capital sums start being sunk into long-term projects which cannot be unscrambled. Manufacturing and trade patterns which cannot be controlled will be determined for years ahead.

Yet energy technologies and innovations are changing almost weekly. A year is becoming an age. Has not his own Government, the Minister reflects, had to change its whole energy strategy over the last three years? How on earth can he pronounce on projects which are going to take ten years to complete and which may be overtaken by events before they start?

No wonder the Minister can hardly bear to look at the submission in front of him.

Nuclear? Sounds wonderfully clean, but probably means a lifetime of constant panic, endless political protest movements and huge claims on the public purse which the Treasury will curse him for, since private investors will just not take the long-term risk. Anyway, will anything actually happen? Didn't a past Planning or Energy Minister 'approve' 12 new giant nuclear stations back in 1980, and how many got built? Answer, one, and that after ten years.

Coal? The very word sends trembles through the Minister, especially if he is a Tory in the UK, conjuring up memories of miners' leader Arthur Scargill, electoral disaster, strikes, years of imports ahead, the objections of the whole carbon brigade and probably yet more public funds to develop carbon sequestration, integrated gasification systems and other promising but as yet unproven technologies.

Gas? That's what they are all talking about and crossed fingers that the horrors of recent winters are not repeated. The pipelines are at last being completed, but will they be filled. What will Russia do next? Can the Continentals be trusted? Will the big new Norwegian fields open up in time? Distant memories are stirred of past Energy Ministers being crucified over rising gas prices. And again, huge political rows ahead over where gas storage caverns will be sited (one right under the Chancellor's constituency) and the route of gas pipelines through national parks. Who decides these plans and why do they always choose the hairiest locations?[47]

Renewables? Again political and environmental rows galore and lots and lots of public funding – to subsidize wind farms, bring on tidal power, solar power (where's the sun?), encourage fashionable 'distributed' or 'embedded' generation – that's power generated in or near every home – and biofuels, from corn, maize, sugar beet and all the rest, wonderful but leading straight into the quagmire of agricultural support and searing farm politics.

Of course the Permanent Secretary has the answer. Make no commitment at all, Minister. No one has the least idea which way fuel prices will go. A return to cheap oil could make any plans look foolish. Best to issue some high-sounding reflections about goals and options, need for flexibility, diversity of sources – that sort of thing. Leave it to the market to decide what to do – which of course, with the market being composed of investors and others who want to see their money back before Doomsday, it never will. Anyway, by the time it becomes apparent that nothing has been done, or the lights go out through lack of foresight and sound power investment, there will be another Minister in place.

Meanwhile the clamour from Parliament, from the media, from Cabinet colleagues, from party officials, not to mention focus groups and PR experts, is to 'do something'. The Minister sighs and wishes he was in another Department. Roll on the reshuffle, as long as it does not shuffle him out.

SEVEN. Global energy statistics are mostly unreliable, statistics about oil production and oil reserves especially so. We may be much nearer the abyss than the figures suggest.
The older man in our earlier Hollywood movie cameo is uneasy when he leaves the cinema, not just because things are obviously not as simple as the film suggests or because his younger friend has a point. The most worrying aspect is that he does not really know what to do next. None of the so-called facts about energy supply, about prices, about prospects, about the best way to turn or where to find economy and reliability, seem to be solid or reliable.

He is right to feel uncertain and unguided. A thick fog hangs over all the energy supply and demand figures with which the public in most countries are being presented. This

The climate hits back. Hurricanes and increasing weather extremes will add to energy insecurity.

does not stop energy experts and policymakers speaking with unnerving certainty and making thunderous ex-cathedra statements about energy. But the truth is that nobody knows for sure how large the reserves, proven or unexplored, really are, how much oil is being produced or what will happen next in oil markets, given the incessant geopolitical uncertainties. Expert forward estimates of the situation over the last three years have often been spectacularly wrong.

Individuals, homeowners and motorists do not know what changes to make, or what next to invest in; businesses do not know what energy-saving or new technology expenditure will pay off. Energy generators and electricity companies are in the dark about what kind of power stations, using what kind of fuels, to start building for tomorrow's needs.

Governments do not know which energy sources or technologies to back – or to tax – or how to get a really reliable system of carbon pricing and taxing off the ground. Military authorities do not know which energy resources to defend or protect, or where they should be deploying scarce manpower. International institutions do not know which political destabilization is going to occur next. Developing and growing economies do not know which energy path to follow as their resources drain away into petrodollars.

The media are not all that well served, either, about energy issues. Journalists and commentators find it hard to decide which information to prioritize and what prominence to give to the cascade of statistics and opinions that flows from 'official sources' or from corporations, from think tanks and universities, and from analysts in banks and other financial institutions. Statistics about oil reserves and oil production are especially cloudy because they rely on reports from sources in oil-producing countries which may well have strong motives for distorting the reality.

There is a strong suspicion that most members of OPEC have been inflating their oil reserve figures in order to gain bigger production quotas. There is an equally strong suspicion that OPEC oil producers may be understating their actual production and export sales in order to appear to be within the laid-down quotas. Yet these are the figures on which authorities like the Paris basedf IEA, or the economists and statisticians hired by the big oil companies and banks, have to rely to pull together their global pictures of where things are headed.

Least of all do the forecasters know what is going to happen to the price of a barrel of oil. This is because the issue goes far wider than economics and because it is the plaything of both geopolitical and climatic events. Any morning a key oil facility in Saudi Arabia can be successfully targeted. Any morning in the hurricane season a storm of special virulence can develop, cutting millions of barrels a day out of world production. Any morning a coup can break out in Nigeria, or new restrictions on the private oil and gas sector can emerge in Russia, or any of the events described above can occur.

So no wonder our older man, or for that matter the typical householder and energy consumer, who has a home to light and keep warm, a car to drive, a workplace to get to, a life to lead, is confused.

Concluding Remarks. The *Danse Macabre*

A scene of major energy volatility and disruption is taking shape around us. This is made all the more dangerous by the vacuum opening up in world order as the Pax Americana withers and instability, disorder and violence breed everywhere, although especially in the countries that happen to be the world's main sources of oil and gas. These conditions are all enemies of safe, cheap and reliable energy supplies and of the stable societies which rely on these supplies. A new international order, or a pattern flexible and resilient enough to cope with constant disorder, is urgently required if the energy security issue is to be resolved.

As the centre of gravity in both the world economy and world power shifts to Asia, where will these new sources of influence and stability be found? Should we be looking, for example, to a new body that has formed itself amongst the Asian powers, the Shanghai Cooperation Organization, for a lead where America has faltered? (The Shanghai Cooperation Organization is a group of powerful Asian nations that have banded together, with China in the lead, to bring their weight to bear on world events, and in particular to develop their own energy 'club' and their own pipeline supply network.[48])

What part will the European Union play, given its ambitions – so far unrealized – to act as a force on the world stage and as a 'counterweight' to America? Or could the 54-nation Commonwealth network, which now embraces six of the world's fastest growing and most technologically advanced economies, play a more forceful role and provide the missing platform for those nations committed to the rule of law and various forms of democracy? (For possible answers, see Chapter Seven.)

What is crystal clear is that the Washingtonian belief in overwhelming force as the means of spreading democracy and 'Western values', and thus stabilizing the world's dangerous regions – thereby ensuring reliable energy supplies – is a deeply flawed strategy. The outcome is the opposite. The world requires new platforms from which to operate and a new diplomacy to deploy from them. Without what Tony Blair, the British Prime Minister has called 'a renaissance of thinking' about these matters, energy security is going to be further reduced and international tensions considerably further increased. This is the *danse macabre* between energy and geopolitics which we allow to continue at our peril.

CHAPTER FOUR

SWIMMING IN OIL: ALMOST UNLIMITED GAS

∽ ∽ ∽

The world is in no way running out. Reserves
are getting larger, not smaller. Geopolitics, transport
difficulties, distribution blockages and global warming
fears are the problems, not energy shortage.

∽ ∽ ∽

First things first. Why does everyone talk about oil in terms of
'barrels'? Answer: oil is measured in barrels because the early
commercial oil producers (in Pennsylvania in the 1850s or
thereabout) poured their oil into barrels for convenience. They
chose the standard barrel available at the time for cheapness as
well as convenience. The standard barrel held 42 gallons or
159 litres.

Is the world now running out? No, there is plenty of oil and
gas around the place. The limitations are profitability and technology, not the actual resource.

That sounds contradictory, given all the talk about the
world running out of oil and about the world production
'peak' identified by M. King Hubbert, the much respected
Shell geologist,[49] as well as the current reality which is a

soaring and volatile crude oil price and thoroughly nervous energy markets.

Yet there can now be no dispute that large oil and gas deposits remain. The giant oilfields of Saudi Arabia, like the Ghawar field, may be running down.[50] But despite the widespread view of oil experts only a few years ago, new giants are turning up.

Start first with 'proven reserves'. This is the oil that can be profitably extracted with existing and tested technologies. The 2006 figure here – which has probably risen since – was 1200 billion barrels. This total consisted of 743 billion barrels in the Arabian Gulf states, 59 billion barrels in North America, 79 billion in Venezuela, 17 billion in Western Europe, 74 billion in Russia and its near neighbours, 55 billion in North Africa, 57 billion in sub-Saharan Africa and 40 billion in Asia.[51]

That's the known, relatively easy-to-extract stuff – enough for about 40 years at present rates. But 'proven reserves' are just the start. At the northern end of the Caspian Sea, in the middle of what had been a nature reserve under Soviet Union rule, geologists in 2000 stumbled on the titanic new Kashagan field, probably the third biggest oilfield on the planet, with reserves estimated conservatively at 30 billion barrels, almost twice the entire reserves in the North Sea. Once up and running, the estimates are that this field alone could produce 1.3 million barrels of oil a day.

Thirty miles east, and two decades earlier, Soviet explorers had come upon the Tengiz field, almost as large. The oil in both are very deep (about 12,000–14,000 feet) and tricky to handle. But there are almost certainly more oil bubbles of this mega-size in and around the Caspian.

Even if account is taken just of this oily Caspian region, where the black liquid oozes out of the ground almost everywhere, it confirms how misleading it can be to claim that 'the world is running out of oil' – except in the sense that hundreds of years hence there is bound to be less than now (although even that may not be true for natural gas, which, according to some scientists, is actually being generated and renewed deep inside the earth's core).

But of course the Caspian region is just one of several where new oil finds are opening up. Under the Arctic ice-cap, to the

A damaged oil pipeline near Basra. Iraq, with the world's third largest oil reserves, has to struggle to keep oil exports flowing.

north of Norway, and also to the north of Russia, billions of tonnes of oil and billions of cubic metres of gas lie waiting. Gazprom's Shtokman field, 300 miles off the Arctic coast in the Barents Sea, is the largest offshore gas reservoir in the world. In the Timon-Pechora region of Murmansk, 1,000 miles to the north east of Moscow, high-quality 'sweet' oil is pouring out while large quantities of gas are burnt off, there being no means of getting the gas out of the area. Meanwhile, BP Amoco is developing a large offshore Alaskan field called Northstar (despite facing the embarrassment of having to close down part of its giant Alaskan Prudhoe Bay field).

In Eastern Siberia gas fields have already been identified which dwarf even the Kashagan discovery. Far-from-wild estimates suggest that Russia has around 70 billion barrels of 'proven reserves' to be extracted, and another 200 billion in likely, but as yet unproven, deposits. This would bring it ahead of Saudi Arabia's 262 billion. Plenty more new oilfields await exploration and development off the coast of Africa. Record prices are being paid for exploration licences for new territories, for instance off the coast of Angola. Madagascar has found more oil. More big oil is probably waiting to be found in Alaska, alongside BP's giant Prudhoe Bay field. There may well be vast reserves round the Falkland Islands. Even tiny Lebanon has oilfields offshore waiting to be opened up – one day, if and when the politics settle down.

In short, while there are many good reasons for reducing oil dependence in the twenty-first century, shortage of the oil resource itself is not one of them. Expert estimates about the so-called ultimate recoverable resource base (i.e. what is known to be there and is extractable with a reasonable margin of profit) have consistently grown over the past few decades. Estimated world oil reserves doubled from 630 billion barrels of oil in 1975 to 1189 billion barrels in 2005.[52] How this fits in with the widespread view that the world is running out of oil, no one can explain.

What makes the estimates go up continuously is a combination of economics and innovation. The IEA explains the process this way: Reserves are being constantly revised in line with new discoveries, changes in prices and technological advances. As oil companies adopt technologies, such as horizontal drilling techniques and computerized reservoir management systems, the estimated recovery rates are being revised. These revisions – always upwards – inevitably add to the reserve base. A few decades ago, the average oil recovery rate from reservoirs was

20 per cent. That is to say about a fifth of the oil known to be sitting in a particular field could be got out, before it all became unmanageably costly and difficult.

Now, thanks to remarkable advances in technology, this has risen to about 35 per cent today. But despite this improvement, two-thirds of the oil known to exist in reservoirs is still abandoned as uneconomic, leaving room for tomorrow's discoveries or innovations to lift recovery rates and magically push Mr Hubbert's peak, the moment when consumption exceeds production plus discovery of new recoverable reserves of oil, ever further into the future.[53] This is why cheerful 'heretics' of the oil world, like Professor Peter Odell of Rotterdam's Erasmus University, point out that the world is 'running into oil' rather than 'out of it'.

Brazil may prove another confirmation of this optimism. The experience of Brazil's offshore drilling is proving that giant new oilfields are out there, waiting to be discovered, just offshore along the continental shelf. Petrobras, Brazil's largest oil company, is moving Brazil from being nearly 100 per cent dependent on foreign oil imports only some 50 years ago, toward becoming a net oil exporter in the next few years. Spectacular results have been achieved by developing the technology of drilling ultra-deep offshore wells in Brazil's Barracuda and Caratingua oilfields, in the Campos basin region.

These two fields are expected to add 30 per cent to the current one million barrels per day of production. Their proven oil reserves are estimated at 1.229 billion barrels. Together they are expected to produce 773 million barrels of oil by 2025. And this is a country which is going to rely on mineral ('rock') oil less and less as ethanol from sugarcane replaces conventional oil in almost every tank. Brazil will therefore shortly become a net exporter of conventional oil, adding substantially to available world supply.

A false picture of the situation has been reinforced in Northern Europe by the widespread belief that North Sea oil and gas are fast running out. The North Sea's role as a major oil and gas source is very far from ending. The talk until recently was of the North Sea running slowly down, but it has already picked up once from apparent decline and could do so again.

One third of its oil and gas reserves, as currently identified, have not yet been produced. The Norwegian side still has substantial gas and oil deposits to develop. On the UK side, the big oil companies may be pulling out in the belief that there are no more giants like the Brent field to be exploited, and are looking for juicier finds elsewhere (with limited success). But for the smaller companies rich pickings remain and the number of wells they are planning to drill stays high. More still lies in deeper northern waters and here, too, the march of technology opens up more accessible and 'economic' (i.e. profitable) oil recovery all the time.

It is true that UK North Sea gas deposits have been depleted very fast in recent years, as the UK electricity industry has switched from burning virtually no gas to using gas to produce 39 per cent of all UK electricity, making the UK Europe's largest market for gas. This, as noted earlier, has caught UK energy planners dangerously off balance, but in the North Sea as a whole, including the Norwegian sector, deposits are increasingly plentiful, with giant new supplies becoming available.

Talk of the North Sea 'running down' is just another misleading generalization in the fuzzy world of oil and gas production statistics. According to the United Kingdom Offshore Operators Association, the UK could still be producing the equivalent of 65 per cent of its total oil requirements in 2020. The 'decommissioning' of wells – that is closing them down and capping them, which some believed would be widespread by now – may well not begin for another 10–15 years. Even in UK waters significant discoveries are still expected in the North Sea, like the discovery of the Buzzard field in 2002. The field contains recoverable reserves of over 400 million barrels – the biggest find in more than a decade.

Other basins, such as the Gulf of Mexico, have undergone similar up-and-down-and-up-again estimates. The Offshore US Gulf of Mexico has become one of the 'hottest' exploration areas in the world, just a few years after it had been declared a 'Dead Sea' for exploration potential. Dramatic improvements in 3-D Seismic technology (showing exactly where the oil lies deep under the seabed) and deepwater drilling methodology largely account for this resurgence. There could still be significant untapped oil and gas reserves, amounting to tens of billions of barrels, hidden well below the Gulf of Mexico. The undiscovered oil and gas potential of the Gulf of Mexico is very large; the deepwater potential for finding more is big, and according to some bigger than the North Sea! Ever-advancing technology will reveal it all in due course. For Chevron it has already done so, with the identification of a massive new field (possibly up to 15 billion barrels of oil) in a province around 175 miles off the Louisiana coast.

Meanwhile, Norway has a potential for maintaining its oil production from the North Sea for another 50 years and its gas production for another 100 profitable years.[54] Estimates say one third of the total petroleum resources on the Norwegian continental shelf are still unexploited. The undiscovered resources in the North Sea are estimated to amount to 7.5 billion barrels of oil equivalent, 4.4 billion of which is liquid (oil, natural gas liquids (NGL) and condensate) and 3.1 billion is gas.

The Russian Energy Empire

Russia is the big one, the world's largest holder of gas reserves and the third-largest oil exporter (after Saudi Arabia and Iran). In the central Moscow offices of TNK-BP, the giant joint venture in which BP has invested $7 billion, the two key executives driving the whole enterprise forward give their version. There could be, they say, as much as five North Seas up in the High North under the ice. Five North Seas! The extraction costs would be high and the technology is not yet good enough. But the oil is unquestionably sitting there, deep beneath the surface. Meanwhile, there is the whole of East Siberia to open up, with its colossal gas fields, as big as the Iranian/Qatari giant gas dome, as well as big oil potential.

Only someone who has never been in Oslo or has never conversed with these Moscow oil men, let alone visited the giant Caspian basin, could ever believe that the world was running out of fossil fuels. Norway and Russia together show that while there are plenty of other problems and dangers with oil and gas, and plenty of arguments for seeking cleaning energy sources, scarcity and imminent run-down are just not the issue.

Russian inclusion in the Group of Eight industrialized nations (G-8) reflects in large measure that country's role as a dominant global energy supplier, giving the country new confidence after the tribulations and humiliations of the Soviet era and the years immediately following it. No one really likes to put it that way, and the thought of Russia making up for lost power and prestige from Communist days by becoming an 'energy power' makes people a bit queasy. But that is the fact of the matter.[55] Since 1998, Russia's oil industry has gone through a significant revival after a catastrophic oil collapse in output in the 1990s.

The country's reserves estimates go up and up. In its annual statistical survey of world energy, BP has recently revised its estimates of Russia's total proven oil reserves to 69.1 billion barrels, 6 per cent of the world's total, up from 45 billion in 2001. But these estimates probably just scratch the surface of Russia's real potential. According to a recent study,[56] Russia's true recoverable reserves are between 150 billion and 200 billion barrels – well into the Middle-East league. That's sharply up from industry estimates of 100 billion barrels a few years ago.

Increasing recoverability (all the clever new engineering methods for extracting oil) and new discoveries of oil account in large part for this upward surge. Early in 2006, Russia's largest private oil company Lukoil announced that it had made a discovery of yet another large oil and gas condensate deposit in the northern part of the Caspian Sea. The field's probable reserves are estimated at 600 million barrels of oil and 34 billion cubic metres of gas.

Russia is easily the world's largest gas producer, with gas fields stretching from Western to Eastern Siberia. The country is already the primary gas supplier to Europe (and becoming a cause of considerable disquiet, of which more in Chapter Five) and in the next two decades it will capture important gas markets in Northeast Asia and South Asia.

No proposition that the world is running out of fossil fuels could stand up for a moment against the basic Russian situation. The problems lie elsewhere – in the geopolitics of the whole region, in the insecurity of long pipelines, in the lack of past investment, and in the seething and underlying tensions across the whole gigantic landmass. Russia could not provide a clearer example of why, with such enormous primary energy resources to hand, there are still so many immediate energy problems.

Look North for Energy

Then there is Norway. Kim Travik's eyes glint. Norway's Deputy Foreign Minister sits in his office beside a large map. The map shows the top of the world, with Norway, Russia, Alaska and Canada all head to head in a neat circle round the polar ice-cap 'sea'. Somehow the world's geography looks different when one looks down on the globe from above. (In Australia and New Zealand schoolchildren are taught from maps putting the South Pole at the centre of everything. Somehow the world looks different to them, too.)

His hand points to the area above where Norway and Russia meet. Here, he says, there are probably a quarter of the entire world's so-far-unexplored reserves. That is 200 billions of barrels of oil – along with almost countless billions of cubic metres of gas, second only to Saudi Arabia.

This, he continues, could change everything. It could change the geopolitics of oil and even more of gas and it could change Europe's frightening dependence both on the Middle East – for oil – and Russia – for gas. It could mean the world will always have plenty of oil, if it wants it – although at a price – and almost unlimited supplies of gas, pipelined or frozen, certainly enough for hundreds of years. Here in the High North and the bleak and frozen Barents Sea lies the future of world energy.

As he speaks the calculations and confident predictions of the past melt away even faster than before – 'the world is running out of oil and gas', 'all the major fields have been discovered', 'dependence on the Middle East can only increase'.

None of this will eventually be so if he is right. It will be vastly costly to extract these huge quantities from under the

Arctic Ice. But then there is technology. People always underestimate the pace at which new technology can bring production costs crashing down. They underestimated its impact in the North Sea, in the Gulf of Mexico and in a dozen other offshore locations. The same will apply in the High North. There will always be oil and always be gas. The cheap-to-produce stuff out of the desert may slow down, or for political reasons it may be interrupted or cut off altogether. Maybe. But at a price the black gold will be available and gas will anyway be plentiful, and convertible into high-quality diesel. Moreover, as the Arctic ice melts with global warming (and it is happening, for whatever reasons) the riches of the region become easier to see, and of course to extract.

Mr Travik was telling us that the scene was about to change – radically and fascinatingly, and in ways which were once again going to wrong-foot the expert establishment.

Geologists put the value of this icy prospect at the mind-boggling sum of 250 billion euros in the Norwegian parts of the Arctic Sea alone, an area of 27.7 million square kilometres. The oil and gas reserves combined may be worth twice that of Norway's existing oil fund, according to calculations by Statoil.

The Barents Sea is the least-explored part of the Norwegian Continental Shelf. In 1980 the Norwegian Parliament decided to open the southern part of the Norwegian sector of the Barents Sea for petroleum activities. Two commercial discoveries have been made – the Snøhvit gas field and the Goliat oilfield. The Snøhvit project is one of Norway's major natural gas projects. It is significant for the fact that it will be Europe's first liquefied natural gas (LNG) export terminal.

The world, as crude oil prices settle on a higher plane than ever before, is getting intensely interested. Top US, European Union, Russian and Norwegian officials all agree that the Barents Sea has the potential to be a new world-class petroleum province and a key energy provider for Europe and the USA – one day. The Barents Sea could offer the USA a stable alternative supply of both crude and liquefied natural gas. The EU could come to look at the Barents Sea in the same way. For the UK in particular, Norway is an obviously attractive and neighbouring source of gas that can and will contribute to the country's security of energy supply.

Similarly for Norway, the UK is an attractive and expanding future gas market that can be used to transport the Norwegian gas supply to other European markets, through the existing pipeline base extending from the UK to other European countries. It works both ways.

On 24 April 2006, Norwegian Foreign Minister Jonas Gahr Støre published, in the Swedish newspaper *Svenska Dagbladet*, a letter, article entitled 'A sea of opportunities'. In that letter Støre urged the EU to look north for energy. Increased extraction of oil and natural gas from the Barents Sea could provide Europe with its much needed energy, the Minister said. The letter sketched a European energy scenario focusing on the ice-packed northern parts of the planet.

The traditional conception of the world, classically defined with Europe in the middle, and America and Asia on the outer wings, is only one way of looking at the map. If instead we place the Nordic areas in the centre, the perspective changes dramatically. The Arctic Ocean is then a mutual sea between Europe, Russia and North America, Støre said.

He might have added that, with the Arctic ice melting fast, about which it is too late to do anything, the whole zone is on the point of becoming increasingly navigable, dramatically reducing sea transport times between Europe and Asia and the Americas.

Inevitably, snags abound and politics intrude. Disagreement between Moscow and Oslo about how to demarcate the border stretching into the Arctic sea has made the 173,000 square-kilometre disputed area – estimated to hold 12 billion barrels of oil equivalent – all but untouchable for exploration and development over the past 30 years. Russia claims the borderline follows the western coherent line. Norway claims it follows the eastern non-coherent line.

The problem comes not just from oil but from fish. The Barents Sea contains one of the world's richest fish stocks. Russian negotiators drew a line from the Russian-Norwegian border on land straight up to the North Pole. With that kind of mapping the Norwegian tip of land at Varangerhalvøya would have turned Russian. This was understandably rejected until recently by the Norwegian negotiators, who argued that the borders at sea must be at the same distance from Russian and Norwegian land territory. Naturally this border must be east of the suggestion from the Russian side.

But in the past year, the stalled energy dialogue between the two countries has thawed as they recognize that their cooperation would help unlock Barents Sea resources. Norway's Statoil thinks there's more oil and gas in the disputed area than in the entire Norwegian sector already set aside, but secrecy abounds. Russia wants the expertise and technology Norway has developed over four decades of working in its own frigid waters. The Norwegians are immensely inventive and have used their energy experience to build a formidable technological lead in energy extraction processes under all conditions. Norway is the world's biggest operator of submarine gas pipelines and of sea-bed extraction and transfer systems, connected to floating platforms which have replaced the giant concrete platforms of the 1970s and 1980s. Today, technology developed on the Norwegian Continental Shelf is setting the pace all over the world.

As everywhere else, the key to it all is cost. At $20 or $30 for a barrel of oil these icy northern areas looked thoroughly uninteresting. At $50 or $60 a barrel they suddenly look seriously attractive to oil companies and investors. Similarly, tapping Russia's vast oil pool will require billions in investment, especially in export pipelines. Unless prices stay high it just will not happen. As always the other key question hangs over the oil scene – are high oil prices here to stay?

The Qatar Cornucopia

Switch now from the chilly north to the boiling desert areas of the Arabian Gulf.[57]

Through the shimmering desert heat the outlines of a strange metallic city take shape – mile after mile of tubes and turrets like the gigantic intestines of a tin man. This is Ras Laffan, the gigantic Qatari complex built on the flat sand to receive ashore and process gas from the great offshore gas dome which the Qataris possess within their waters. James Baldwin, the young but seasoned site manager, explains over canteen lunch how the gigantic operation works, as gas is filtered, cleaned and then frozen into a liquid at −162°C, where it is at one ninth its volume in gas form, then pumped into specially designed ships and sent off to the markets of the west and of Asia. The whole process is not cheap, but is a much cleaner product than oil and so far very reliable.

The Qataris have taken a gamble. But it is a shrewd one. They have gambled that both frozen gas (LNG) and gas turned into liquids (GTL) – mostly diesel – will be competitive with pipeline gas and will be a crucial part of the energy supply mix, especially for Northern Europe, for Japan and for the United States. Their gamble looks like paying off.

Abdullah Al Attiyah, Qatar's long-serving Energy Minister, sits in his modest office and muses on the position in which his country now finds itself. At first, he says, the Qataris were disappointed that their 'legacy' was a vast dome of gas in which nobody had much interest. Gas used to be the unwanted orphan fuel, as often as not burnt off as it emerged with oil. A gas field discovery used to be called a dry hole.

Now the scene has changed radically. The tiny state of Qatar (population around 500,000 – the same as Bristol in the UK – although only about 150,000 are Qatari citizens) found that it was sitting on an energy cornucopia. The soaring oil price, and the flutter of apprehensions now passing through European capitals about the reliability of Russian gas pipeline supplies, have played into its hands. Suddenly frozen gas can be called up by energy-hungry consumer countries to fill the gaps, back up pipeline supplies and meet moments of high demand or crisis.

Soon it will do more than that. Frozen gas needs big terminals to receive it, warm it up and feed it into national and local gas grids.

Not content with existing LNG terminals in Northern Europe, notably the BP-owned one at the Isle of Grain in the mouth of the Thames, or with another giant terminal being built by ExxonMobil, the Qataris are building their own 80

Not just a man's world. Co-author Carole Nakhle on the Gulfacks C North Sea oil platform, in the Norwegian sector, with Statoil operatives.

per cent-owned LNG terminal at Milford Haven in South Wales that will supply LNG regularly and on long-term contracts (hence a more secure source of supply). This will be gas not only for the British market (and fully enough to meet most needs when combined with Norwegian sources), but the route can also be used to re-export to Continental Europe and diversify Europe's sources of gas supplies in a way that government planners and strategists have so signally failed to do for themselves.

LNG can replace oil almost completely – in theory. The problem is the cost of processing and transporting it. There is no prospect for decades ahead of shortage; it produces about 50 per cent less carbon dioxide than oil and it can be drawn from a variety of different areas. It could thus be an important element in building up the diversity pattern which energy planners now keep urging, but which somehow got overlooked by the EU during the years of rising dependence on Russian supplies. It was Winston Churchill who preached diversity eloquently at an early stage when, back in 1913, he ordered the vast British fleet to change over from coal firing to oil firing. Challenged about relying too much on the Middle East for supplies, he insisted that 'safety and certainty in oil lie in diversity and diversity alone'.[58]

The world may still be swimming in both oil and gas but the problems lie in getting the stuff out – that is to say with accessibility, with overcoming both climatic and local political conditions and with transporting the black fluid, or the gas, from remote and often bleak spots to the refineries and markets where it is needed. All the time this is adding formidably to the cost and the risks. There was a time, back in the last century, when all the new oil finds seemed to be in the middle of flat deserts in politically stable regions with blue calm waters, near comfortable shipping channels and port facilities. That was once the reality, but today in the Middle East it is 'such things as dreams are made of'.

But how long term is all this? Will it help ease shorter-term energy security pressures that are building up in gas markets and which have already seen gas prices rocket upwards?

Frozen gas is still only a small part of the current gas supply scene, although it could become much larger very quickly. About 5 per cent of Western Europe's gas imports are of the frozen variety and the IEA estimates that this could rise to 25 per cent by 2020; 10 per cent of US gas imports are frozen (gas imports in turn being about 30 per cent of total US gas consumption each day).

For the big international oil companies – the 'majors' – all this is distinctly uncomfortable. In country after country their access to new oil is being shut off by nationalistic policies and state-owned oil companies. This adds to the general air of pessimism. From the big companies' point of view the oil outlook is inevitably gloomy. There may be plenty of oil to come, but they cannot get at it. Overall, oil and gas companies that are either directly or indirectly state-run now control around 85 per cent of all production and known reserves. National champions in Russia and China are racing to corner assets all round the world. Life for the old guard, the long-established companies like BP and Shell, gets harder and harder. Their reserves are falling despite ever greater expenditure being devoted to exploration.[59]

As Professor Peter Odell of Erasmus University keeps warning the big oil companies – and his message is not very welcome – there is plenty of oil in the world and recoverable or 'proven' reserves keep expanding. But to survive and prosper, the great international energy companies will have to reinvent themselves, as BP is already trying to do by claiming that its logo – BP – now stands not for British Petroleum but for Beyond Petroleum.[60] From the international oil companies' (IOC's) viewpoint this is the wise, indeed the only, course. They – that is the four remaining oil giants out of the original Seven Sisters[61] – have a major part to play in changing public understanding, habits and demands and then meeting those demands swiftly and efficiently (and profitably).

Conclusion

The era of oil discovery is plainly by no means over yet. Russia, the Norwegian High North, West Africa, the Caspian, the ultra-deep waters off Brazil and even the heavily explored Middle East abound with oil – and with gas deposits. Iraq has over 130 un-drilled prospects.[62] Early in 2006, Saudi Arabia announced a colossal new gas find in its oil-rich Eastern Province.[63]

There is something askew in the whole debate about energy supplies. In no way is the world running short of oil (or gas). That is simply not the issue. Almost all the problems lie not in the resources themselves in the ground but in the formidable and ever-growing risks and costs of safeguarding them (which may prove the biggest cost of all and not succeed anyway), in transporting them to market and to the points of distribution and consumption, and doing so continuously, efficiently and reasonably cheaply. In the way stand threats and roadblocks of a kind which oilmen and gas technicians never had to face before, and which bewilder policymakers and confound the efforts of diplomats and statesmen.

If these issues cannot be resolved – some can, but some of them are truly intractable – then it is either substitution or saving energy, either alternative energy sources or much greater economy and efficiency in the end use of oil and oil products, which will have to take the strain – probably both. Both the alternative sources of supply and the ways of curbing demand growth are largely there and ready to be deployed – and should have been long ago. By standing a little higher and peering forward it could have been discerned that the new supply threats were coming. But no one on the watch gave the cry and the world now faces the immediate and potentially devastating consequences.

CHAPTER FIVE

THE PIPES OF WAR

∽∽ ∽∽ ∽∽

Pipeline Dramas: Gas, politics and power in Central
Asia. How Europe became in thrall to Russia.
Gas is the wobbly stepping stone to the energy future.

∽∽ ∽∽ ∽∽

For Europe reliable energy now means reliable flows of piped
gas: hourly, daily, continuously and at the right pressures.
Once a supremely boring and technical subject, pipeline gas is
now centre stage and has accordingly developed its own dramas and rivalries, some of them well up to the level of intrigue
and violence as depicted in James Bond movies.[64]

These ensure that while gas may be the next stepping stone
on the route to a safer energy future, which the carbon campaigners are silly to try to sidestep, it will prove, at least in the
European case, to be a distinctly wobbly one. While the exaggerations about running out may be fading, as technology and
rapid innovation enable more and more massive deposits of
both oil and gas, the dramas about transmitting and delivering
them to an ever hungrier and larger market are only just
beginning.

A Dinner in Belgravia

The dinner party was going well. At the long table in the first-floor dining room of the German Embassy in Belgrave Square, London, a touching celebration was taking place – a tribute by the Federal German Government to the sculptor Henry Moore, whose works had become widely admired in Germany.

Present were the German Chancellor, Helmut Schmidt, visiting London for the occasion, the British Prime Minister, Margaret Thatcher, the German Energy and Industry Minister, Graf Otto Lambsdorff, and his British opposite number, David Howell. The date was 23 October 1980.

The conversation was turning to energy, despite the artistic theme of the evening, because the world was preoccupied with the subject. Suddenly there was an interruption. Margaret Thatcher, glaring down the table like a teacher who has caught an unauthorized aside at the back of the classroom, was asking for something said by Helmut Schmidt to be repeated – slowly.

What Schmidt had said was that West Germany (this was years before the fall of the Wall and reunification) now relied upon the Soviet Union for 14 per cent of its daily natural gas consumption.

The Thatcher coffee cup came down with a crash – 14 per cent! That, she said, was very dangerous and unwise. How had it been allowed? What was Helmut going to do about it? Schmidt replied in the slightly patronizing tone he still (mistakenly) used when talking to the English lady. 'My dear Margaret', he said, 'the Russians have always been the most reliable suppliers. They need us as much as we need them. There is no danger at all'. The talk moved on.

On 1 January 2006, just over 25 years later, Europe woke up to some disturbing news. Russia was playing politics with gas and the repercussions were being felt all over the European system. Russian policy had been to raise gas prices sharply to foreign markets, after years of exporting gas at highly subsidized rates to its satellites and friends. If Ukraine was no longer a friend, let alone a Soviet satellite, then it would have to pay the market price like other people. The Ukrainians responded by diverting gas intended for Western European markets, and the pressure in those markets sagged accordingly.

So who had been right and who had been wrong back on that evening for Henry Moore at the German Embassy? Today, through its state-owned monopoly Gazprom, Russia alone accounts for over 50 per cent of European gas needs. All Russian gas exports to Europe (except deliveries to Finland and the portion of Turkish exports delivered via the Blue Stream pipeline) transit through three countries: Ukraine, Belarus and Moldova. The map on page xx shows the present dense network of pipes, with another one under construction across the Baltic.

For years all had seemed well. The reassuring Schmidt view was apparently vindicated. The Communists were thoroughly reliable people with whom to do such business. Right through the fall of the Berlin Wall, through the Gorbachev era and his unseating, through the Yeltsin period into the Putin era, the gas kept flowing in growing quantities.

The sudden fall in volumes delivered to European Union countries caused an outcry all over Europe – and an extreme attack of nerves. By 2 January 2006, Hungary was reported to have lost up to 40 per cent of its Russian supplies; Austrian, Slovakian and Romanian supplies were said to be down by one third, France by 25–30 per cent and Poland by 14 per cent. Italy reported having lost 32 million cubic metres, around 25 per cent of deliveries, during 1–3 January. German deliveries were also affected.

On 2 January 2006, Gazprom reacted by saying that it would pump an additional 95 million cubic metres per day into the network to compensate for Ukrainian withdrawals. By 3 January, Austrian and Hungarian supplies were back to normal levels, although some other countries were still experiencing shortfalls. By 4 January, Russian gas deliveries to Europe were back to normal levels.

As it turned out, despite the uproar, no EU country actually needed to interrupt supplies to customers as a result of the unwelcome reduction from the East. The position was certainly made easier due to relatively mild weather in Europe for the time of year and the fact that many commercial and industrial customers were not operating over the New Year holiday period. But brows perspired and memories went back to the Margaret Thatcher warning all those years ago.

A Muddled European Response

The Ukraine dispute lasted a mere four days, three of which had resulted in shortfalls to European supplies. Oil shocks back in the 1970s and 1980s spread over months. But this time four days were sufficient to send Europe's energy planners into a spiral of anxiety, reinforced by its coincidence with yet another bout of rapidly rising oil prices and the announcement of big retail gas price increases to follow.

The debate duly mushroomed into an argument about European energy policy (was there one?) and a predictable call for a new and common policy for the whole European Union. Amidst the noise of clanging stable doors it was felt that the concerns about use by Russia of its vast oil and gas reserves for political purposes, plus the global surge in oil prices, all pointed to the need for EU governments to pursue fundamental changes to European energy policy in the hopes of reducing dependence on Russian energy in general.

Liberalization was going to be the key, combined with pressure on the Russians to open their monopoly gas pipelines to other countries and other suppliers.

The late-in-the-day realism of all this looked, and still looks, doubtful. Official predictions show the EU's gas import dependence on Russia, far from falling, could grow to 70 per cent of general energy consumption by 2030, exposing Europe to much greater risks than now from those 'economic pressures' which so worried the EU offficials.

Every story has two sides and the saga of Russian gas supplies to Ukraine does need understanding in its proper context. Once Ukraine and other former parts of the Soviet Union had become sovereign states, it was inevitable that Russia would abandon the highly subsidized domestic price structure of the Soviet past and start pushing prices for their fuel exports to market levels.

The Ukraine event was part of a long chain of gas supply arguments between Russia and its neighbours (and former satrapies), which had been brewing up over the preceding year. There had already been a bust up with Turkmenistan over gas prices and in July 2005 the Russian Duma voted unanimously that CIS countries[65] – Georgia, Moldova, Ukraine, Estonia, Latvia and Lithuania – should pay 'world' (i.e. European) prices for gas.

So by 2006 there were fertile grounds for yet another gas row – this time a big one. But Europe simply failed to see the warning signs and reeled in confusion about what on earth could have gone wrong.

A policy paper seemed the answer and one duly appeared entitled 'Secure, Competitive and Sustainable Energy Policy for Europe'. This called for steps to diversify EU supply 'by fuel, by source and by supply route', establishing ways to intervene if specific EU nations face a energy crises and acting together when addressing the rest of the world on energy issues. Hopefully, it urged that EU countries should 'speak with the same voice' on energy issues, integrate gas and electricity grids, diversify fuel supplies and lead the world in energy savings.

In addition to an already successful [*sic*] EU 'dialogue' with Russia, the paper suggested adopting cooperation pacts with 'other key producer and transit partners of the EU, notably its Eastern neighbours, the Caspian Basin, Central Asia, Southern Mediterranean, the Middle East and the Gulf region'. More talks with Norway were called for and a better dialogue – the old cry – with the OPEC producers.

Overall, it was asserted, a 'road map' would be needed to reduce Europe's dependence on imported oil by way of energy efficiency in the transport sector and using different types of fuel. EU oil stocks should be made more readily available for emergency release and more transparent by regular publication of stock levels – a task already agreed within the IEA; 973 billion dollars of plant and infrastructure investments would be needed over the next 20 years. EU grants and loans should have more focus on energy security. Connections between EU countries for gas and electricity had to be improved. 'For a real European electricity and gas market to develop, the electricity and gas grids have to function as European grids', it said. At the same time, the Commission said it would start antitrust probes to open gas and electricity markets to more competition.

More competition, and therefore, hopefully, lower prices, is desirable in itself. But if the core problem is overdependence on one source and the need to diversify European energy supplies, how is the competition goal meant to help? This remains a mystery. Like many other 'official' reports portending to wrap problems in an EU envelope, this one tried to paper over contradictions and present the policies proposed as harmonious, compatible, suitably *communitaire* and practically feasible. But the contradictions were, and remain, glaring. As usual the Brussels longing for more centralization is rubbing against the desires of member states to handle their own distinctive situations separately. As Germany's Energy State Secretary observed:

> We share the diagnosis of Mr Barroso (EU Commission President) that the market is not working at the moment, but the problems in the electricity market differ greatly at national levels. We do think decentralised control is more effective. *Nobody wants a mammoth European body.*

If these anxieties were not enough, the summer of 2006 brought a new nightmare to Europe. Russia and Algeria, the two main suppliers of gas to Western Europe, announced plans to work together, not just on technical matters but in marketing arrangements as well. The implications were clear. The two big suppliers wanted to use their strength. The Italians felt especially worried. Of their large volumes of imported gas, the Algerians, through their state-owned company Sonatrach, supplied 37 per cent. The Russians, through state-owned Gazprom, supplied 32 per cent. Together they were a pair of pincers, squeezing the Italian market.

Pierluigi Bersani, the Italian Industry Minister, dashed off an anxious letter to the EU Commission in Brussels, pointing out that the threat now hung over the whole EU and that something must be done. The Commission concurred, something had indeed to be done.

At the end of 2006 there was yet another panic as the Russians exerted their energy muscles again, this time in a dispute with neighbouring Belarus over both gas and oil prices. Accusations flew that Belarus was siphoning off oil from transit pipelines destined for western European customers. Flows were temporarily cut and then restored.

There will be more such incidents as Russia strives to move its former satellites and semi-autonomous states on to an arms-length customer basis. And the repercussions will undoubtedly be felt from time to time in Western Europe.

So, yes, all are agreed that steps must be taken and the Brussels air is full ideas for more diversity, more coal burning, more nuclear power and more energy efficiency. But is it all too late? Perhaps the EU planners should have thought of this all those years ago, when Margaret Thatcher gave her warning. The wait for Europe to speak with one voice on energy issues may prove a very long one indeed.

The core dilemma facing the would-be EU energy policy architects is still the same one which faced governments all the way through from the 1980s. Powerful investment in much greater energy efficiency and alternative sources (and therefore security) required the 'reassurance' of lasting high prices for oil. Once there was that reassurance, then all the alternatives, all the unconventional sources, all the energy-saving technologies would begin to look worthwhile.

But that was not the signal that the markets were giving through most of the last two decades of the twentieth century and certainly not what the public wanted to hear or welcome. Clearly an oil price held for certain, and for a prolonged period of years, at more than $100 a barrel would create a massive economic incentive which neither the current subsidy nor tax policies could ever generate.

This of course would involve turning desultory 'dialogue' between producers and consumers, both with a deep common interest in security both of supply and of demand, into something ten times more positive and concrete.[66]

Unsurprisingly, neither the EU Commission Energy Report, nor the consequential new European energy 'Plan' to which it in due course gave birth, aired any of these heretical and dangerous thoughts. They were, and remain, taboo. Both the paper and subsequent announcements, which followed at the end of 2006, painted an illusionary picture of harmony amongst the member states of the EU. There is none.

Meanwhile, the EU's energy planners may be missing a much more dangerous challenge directly ahead of them. Russia's own domestic gas consumption is gloriously wasteful – gas being heavily underpriced for political reasons. The strong likelihood is that within five years Russia will either have to take drastic action to cut domestic demand, at heaven knows what political price, or it will fail to meet its export obligations. The investment needed in new gas fields both to satisfy domestic Russian consumers and to meet export targets has just not happened.

The consequence is that unless Russia adopts more sensible gas tariffs at home, and unless it moves to a very much closer degree of cooperation and mutual investment with western European gas companies and distributors, Western Europe will face serious gas deficiency problems. Russian gas industry leaders insist that their customers will always come first. But in practice, without far more speedy and intimate cooperation between Russia and Europe, we can look forward to crisis-strewn winters in Europe and plenty of recrimination all round.

Behind the immediate panic about Russian gas reliability lie even deeper worries about Central Asia and its key role in global energy supply.

The Caspian basin is now believed to hold more oil than the entire North Sea. The key question is, as it always has been since the earliest days of oil extraction in the region,[67] how to move the area's oil and gas to world markets. Every pipeline route raises new dangers and stirs old rivalries. Azerbaijan, at the centre of the Caspian oil region, has been locked in a deadly struggle with its Armenian neighbour (over Nagorno-Karabakh), so western routes have to be planned to somehow bypass Armenian territory. Georgia, with its seemingly unending record of civil wars, massive corruption, breakaway sub-regions[68] and general chaos, is seen as 'the linchpin country for the export of oil and gas to western countries'.[69] Russia hovers around, dubiously involved and clearly longing to regain control of the region. North of Azerbaijan sits blood-soaked Chechnya, again a key route for Western pipelines. Indeed, Grozny in Soviet times, and until it was flattened by Russian forces in the 1990s, was a hub for Caspian oil transit through pipelines to the Black Sea. A senior Russian observer remarks 'All the Caucasian wars are at least partly about oil'.

The Americans certainly see Caspian oil as part of a new pattern of more secure oil supply to world markets, the idea being to rely less on Saudi Arabian and other shaky Middle-East sources. Heavy backing was given from the start to the giant new pipeline from Azerbaijan's capital, Baku, on the Caspian, across Georgia to Ceyhan on the Turkish Mediterranean coast. This pipeline, now in operation, must be one of the most politically loaded oil projects of the modern world. America has backed it because it avoids Russian territory – and of course for the same reason Russia has viewed the whole undertaking with deep hostility from the start. The pipeline carefully skirts Armenia, in an enormous loop, because the Azeris hate Armenians and trust them not at all as a transit country. (Have a look at the map of the Caucasus on page xviii, to see how this jigsaw fits together.)

The pipeline through Turkey is also an alternative to another obvious route southwards to the Persian Gulf through Iran. But America with its present policy and mindset ('those who are not with us are against us'),[70] and its refusal to seek any kind of balanced bargain or deal with Iran, will go to any lengths to avoid a line that way as well. A further alternative, now abandoned, was to take the oil out eastwards through Afghanistan, now a cauldron of danger and uncertainty, and no place for pipelines, whether on the surface or buried. Sabotage in every case is easy, and even the Baku-Ceyhan line, supposedly the 'safest' route, has already been milked by Georgian locals.

Meanwhile, the Russians, in a kind of chess game countermove to American tactics, are pushing for a new pipeline from the Black Sea, where it would be fed by Russian tankers, running down to the Aegean, ready to supply Greece, Bulgaria and the rest of the Balkans. Tankers would load up at Novorossiisk up beyond the Crimea, drinking from pipelines coming direct from the Caspian region, and then cross the Black Sea to the Bulgarian side. Mr Putin has already been in Athens tying up the project.

The Afghanistan option raises another aspect of the pipeline debate. In the words of Kazakhstan's chief oil pipeline planner: 'In general we do not want to pump our oil to the West, but to the East, where the hungry markets are'.[71] A pipeline from the giant Kashagan field through to Xinjiang, and from there eastwards eventually to Shanghai, would be greatly to the Chinese liking, although the Japanese would certainly want a look-in, and are already engaged in intricate manoeuvres to get other Russian pipelines to go in their direction. The US Energy Information Administration believes that by 2020 China will consume 10.5 million barrels of oil a day, seven million of it imported – from somewhere. This is not the kind of dependence on foreign sources the Beijing leaders would like, but to avoid it will require huge policy changes – something to which Beijing currently gives a far lower priority than the overriding one of fuelling Chinese phenomenal economic growth. So for the time being the Chinese Government, and their oil agency companies such as CNOOC, are taking no chances and going all out for 'secure' oil and gas supply sources, via pipeline and sea routes, from every possible source – Iran and Kazakhstan very much included.

The best near-term guarantee of oil and gas security in Europe from the eastern direction would be an open set of pipelines from Iran, Uzbekistan, Turkmenistan and Kazakhstan, using, where appropriate, Gazprom's enormous pipeline network. The question is whether this is remotely feasible? Can a friendly dialogue with Russia, or with anybody, achieve this? Experience suggests that negotiating with Russia is possible but requires more than friendliness at the best of times. But Russia's bargaining power has obviously received a further boost from its eager Eastern customers besides Europe. Russia is now increasingly able to deliver energy to either the EU or to China, or to India or to Japan. Russia's eagerness to look eastwards is increasing all the time. President Putin has now spoken in support of a new target for Eastern customers. This is that Russia should aim to supply 30 per cent of its exports to the East. At a time when rumours abound that Russia may not be able by 2010 to fulfil even its current gas contracts with Western Europe (as explained above), this can only add to general European nervousness about the energy future.

The situation is all part of a shift in the tectonic plates of world power, both economic and political, towards Asia, which Western policymakers have been dismally slow to grasp. Could the EU ever have bargaining chips to get results – e.g. veto-ing Russian accession to the WTO – unless the Russians permit freedom of transit through Gazprom pipelines from Turkmen-istan etc.? Or is it all too late and is Europe anyway on the wrong track in trying, sometimes vaingloriously, to act as a world power bloc, a new United States of Europe? The issue goes to the heart of the perceptions which govern the current Western world view – perceptions which are increasingly being challenged as the policies built on them in the Middle East, in central Asia and in the Pacific region seem to falter and fail.

Russian statements on energy issues, and the Russian atti-tude to the proposed Energy Charter,[72] which would open up Gazprom monopoly pipelines to other suppliers to Western Europe, suggest that the going is going to be very slow and very difficult.

According to Valery Yazev, the chairman of the Russian State Duma Committee on Energy, the whole Western approach is flawed. By suggesting that use of Gazprom pipelines should be opened up to auction, the way would be open for cash-rich Western concerns to muscle in. Russia, claims Mr Yazev, would then become a mere 'transit gas corridor'.

The heavy hint in all this is that if Russia is to open up and allow other gas to flow through, then Gazprom will want the same access to Western gas consumers through distribution systems and retail outlets. In other words Gazprom, with its own vast resources, wants to do some muscling-in as well. It is on the move in this direction, having already bought one small British gas company and started bidding moves for a big one – Centrica – which supplies 11 million homes.

Meanwhile, Gazprom sees itself as much more than a monopoly supplier. Its Nordtream gas link, and associated deal with Berlin interests, brings it into direct alliance with German ambitions to supply gas under the Baltic Sea direct to Germany. The EU had high hopes of bringing gas direct from the Caspian region to Austria and Western Europe without Russian involvement or crossing Russian territory, via the 'Nabucco' pipeline. But almost as in a Tom and Jerry cartoon, the Russians have made their counter-move. It is for Gazprom to get there first by joining the Austrians in running the project.

The Russians are indeed coming, and the east is moving westwards. Europe, with all its glorious culture and history, has misread the new world and still struggles to be a self-protected power bloc, when blocs are being replaced by networks and when power has taken a new shape which older generation Europeans, and Americans, do not fully understand.

Eastern Energy Takes Over

The Russia-China supplier-customer relationship is coming along fast.

China and Russia may be uneasy neighbours in the Far East, but they entered a strategic embrace of sorts in March 2006 when Russian President Vladimir Putin unveiled a clutch of energy deals in Beijing. During the visit, Gazprom announced in Beijing that it would build not one but two pipelines, from west and east Siberia, to supply about 80 million cubic metres of natural gas annually to China by 2011. A Gazprom field is expected to supply the western Siberian route. This is tentatively tied to the existing west-to-east pipeline built by PetroChina, the listed unit of China National Petroleum Corp. The eastern route is expected to be supplied at least in part by the Kovykta field.

TNK-BP, the Anglo-Russian joint venture in which BP has invested £7 billion, wants to be in on this scheme as well. It has had a decade-long struggle to develop its giant Kovykta gas field in eastern Siberia, for sales to China. What was announced during President Putin's visit to Beijing in March 2006 was a further step in a dance of delicate complexity and subtlety. In September 2005, Alexander Medevedev, head of Gazexport, announced that Gazprom is pursuing two gas supply options, one from West Siberia to Xinjiang Province and the other from East Siberia and Sakhalin to north China. By 2011, he believed, gas would be flowing.

This plan could be a shade unrealistic, and anyway has somehow to dovetail – or not, as the case may be – with the ambitions of Kazakhstan and the rivalries of Japan and China on the oil scramble. But it could definitely be the shape of things to come. The resources are ample and plentiful, but as always their passage to the points of need and consumption is fraught with tensions and hazards.

In particular, the Russian authorities have continued to make life uncomfortable for major foreign investors – for example, in both the big Sakhalin projects, involving both Royal Dutch Shell and ExxonMobil (and Japanese companies, Mitsui and Mitsubishi), and in the Kovykta field where BP had such high hopes. The Japanese have been especially dismayed at the apparent blocking of huge gas developments in the region, on which their future hopes for LNG supplies were relying heavily. China was supposed to be the Japanese worry. Russia was going to be the no-problems one. But things are turning out differently. As always, spotting the resource has been the easy part, developing and extracting it a political nightmare.

In the evolving 'struggle' for demand/supply diversification between the EU and Russia, another new player has emerged. This is Algeria, a country with its littoral almost in Europe's forecourt and with its huge hinterland, stretching far down into central Africa, loaded with enormous gas and oil deposits.

Since the early 1980s Algeria has been trying to find a political balance which prevents it sliding into fanatical – and murderous – extreme Islamic hands, which restores its economy, shattered by years of incompetent and misguided socialist planning, and re-positions it as a powerful player in energy markets and geopolitics generally. Its hour may now at last be coming, even though the extremist threat remains severe and the ripples from the tensions and terrors further east in the Islamic world spread along through the Mahgreb and along the North African shore.

In March 2006, President Putin visited Algeria and made some interesting deals. These have now been further firmed up,

Endlessly vulnerable. Thousands of miles of overland piped gas and oil present easy targets for sabotage and terrorist attacks.

as we have seen, into detailed cooperation agreements, much to the alarm of EU gas consumers, especially the Italians. The pincers could be closing not just on agitated Italy but on the whole of Western Europe, but for one escape route – Norway.

One of the continuing puzzles in the whole of the EU's official attitude as represented by the Commission in Brussels is why it has paid so little attention to the Norwegian potential for helping Western Europe improve its energy security. Only very recently have Brussels officials, stung into action by the Ukraine drama, opened serious talks with Norway about how it might help fill the breach, as it certainly could. Norway already meets about 15 per cent of EU oil demand through its pipeline system and about the same percentage of EU gas needs. Gas from the huge Snøhvit field in the Arctic could provide another 7–9 per cent.

Has the delay until now in focusing on this obvious alternative to the Russians something to do with Norway's refusal to sign up to EU integrationist ambitions and the constant affront to Europhile pretensions which Norway's super-prosperous economy presents – in contrast to the stagnation and sclerosis which continues to infect the Continental mainland? One wonders. One EU member state, the Czech Republic, has gone ahead on its own and signed a new supply agreement direct with the Norwegians. It is clearly tired of waiting for Brussels. Germany, meanwhile, is enthusiastically combining with the Russians to build a new North European gas pipeline under the Baltic Sea, thus bypassing potentially awkward transit countries – although making Germany not less but still more dependent on Russian energy. Somehow rhetoric and real-life actions do not seem to be pulling the same way.

Norway's eyes are set in particular on the UK as a highly attractive market. This is the one place that appears really good to Statoil executives as they look out across the North Sea from their Stavanger offices, gas prices being determined by open competition, and British households and factories being hungrier than ever as the supplies from the UK side of the North Sea have been drained down rapidly.

Unfortunately UK policy planners, like those in Brussels, have been looking too much the other way until very recently. The UK idea, as local North Sea gas supplies have run down, has been to join in the Continental grid system drawing its gas from Russia. The plan is, or was, that gas from the Continent would flow through a giant interconnector pipeline, attracted by the juicily high prices and free markets offered by the liberalized British supply system.

So confident were the planners that this was the best future that construction of pipelines from reliable neighbouring Norway was allowed to flag. Official UK forecasts for gas imports show volumes from Norway falling away after 2008 and gas from Continental Europe (mainly Russian through the Belgian pipeline connection) rising. In practice the opposite is more likely to happen. The 750-mile Langeled pipeline, hopefully bringing massive flows of gas from Norway's new northern fields to the UK, is at last starting up, but it has been too late to help UK gas supplies in the winters of 2005 or 2006. It will not be filled with the gas volumes the UK requires until Norway's gigantic new Ormen Lange field is fully opened up and operational, which could still be two years away.

Whether in the meantime it will bring enough supplies to head off shortages during the two or three winters ahead remains to be seen. Immediate rescue will probably depend not on the Norwegians but on the flow of gas from the Continental systems.

And rescue has turned out to be what is needed. In the cold spells of the winter of 2005–6 – mercifully rather few – the gas that was supposed to flow from the Continent never came, with the pipeline running two-thirds empty. What the planners had forgotten was that both the French and the German gas distributors were monopolies, acting under strong and direct political influence. That influence was clear – and originated at the highest level. It was to ensure, whatever the market said, that French and German consumers got priority – especially on cold days and even more especially if there were signs of a weakened flow from the Russian direction.

What the planners had also forgotten, or never thought about, was that as UK North Sea gas supplies ran down, large new gas storage facilities would be required and that these would take some years to plan and build – not least because of people's aversion to having gas stored in their neighbourhood.

The net result for British households has been far from pleasant. Gas prices soared to extraordinary levels, violently oscillating during their climb. Over one weekend in March 2006, UK wholesale gas prices quadrupled. Talk of sustainable energy strategies and the better low-carbon world that lay ahead were not much comfort. Later in 2006 the whole system swerved the other way, with so much gas available on warm autumn days in Northern Europe that gas producers were having to pay to get their pipes cleared.

No energy supply system should run on this basis. Gas is plentiful in Northern Europe and will be more so in the near future. But the infrastructure of both pipelines and storage is a 'must'. The UK should be in one of the best positions in Europe, not the worst, when it comes to gas supplies. But this requires forethought and a readiness by government to encourage the timely construction of facilities to handle it. We are not there yet. Energy planners who spend too much time thinking about low-carbon targets in 2050 may not be able to solve gas needs in 2009–10.

The Near Future is Not Oil, it is Gas

If gas in its normal form raises scares and worries then the good news is that a new energy world, although still gas-based, could be just ahead. It is frozen, flexible, more secure, notably cleaner than oil and coming from a big variety of sources – some of them, but not all, highly reliable in a way that oil sources and pipeline gas sources are not.

This source is LNG. It is not particularly cheap and like all transported energy it requires complex infrastructure installations, although it involves no long stretches of unguarded pipelines across hostile and remote territories, waiting to be sabotaged. It can also be made into high-quality and relatively clean diesel for vehicles.

It does, however, have its own security problems. A frozen gas transporter is a hugely vulnerable vessel and a juicy terrorist target. Naval escort protection is often necessary. That is one of the hidden costs of energy which consumers will find they are meeting, whether they know it or not.

Taking frozen (LNG) and pipeline gas together, there is absolutely no shortage of this resource round the world, ready to be pumped, frozen and shipped, or converted to diesel (through the so-called GTL – Gas-to-Liquids – technology). There is no conceivable Hubbert's Peak for gas because no one has begun to get the measure of vast new gas fields to be opened up. These are sited in gargantuan volumes not just in high-risk areas like Iran or offshore Saudi Arabia or the huge hinterland deserts of Algeria, but in middle-risk regions like Eastern Siberia, where probably the biggest gas fields in the world have yet to be opened up, in tiny but gas-rich Qatar, as described earlier, and in low-risk, utterly reliable areas like the Norwegian North Sea and the Norwegian sectors of the Barents Sea and the Arctic ice-cap zone.

Conclusion

Almost all the current and near-term problems of energy security could be, or could have been, resolved by natural gas.

Oil is plentiful but its unreliability increases all the time, as do costs and risks of extraction and protection. Gas sources are more diverse around the world. But gas has its growing problems, too, as described above. Furthermore, although cleaner by far than oil, it remains a fossil fuel so that minds focused solely on the low-carbon future tend to give it a low priority, even as a stepping stone to better things. Key projects for new gas pipelines in the North Sea, and for storage facilities for gas, which should have been developed earlier, are now belatedly coming into the frame. For example, the UK Energy Minister was still 'calling' for more gas storage in the UK as late as June 2006. Developing new gas storage on land (a large new 'cavern' for storage is planned in the middle of Cheshire) could take years and endless planning wrangles.

Somehow in the past these matters did not seem important to the policymakers and politicians. Back in 1981, Margaret Thatcher's Cabinet turned down the proposal from the then British Gas Corporation, and its powerful chairman, Sir Denis Rooke, for a vast new gas-gathering pipeline to bring gas from the emerging new Norwegian fields. For the British Treasury it all seemed too long term to understand. Had it gone forward, much grief would have been avoided in later years. Treasury 'economies' often turn out to be very expensive.

But for the next decade gas will be Europe's key energy source. It will also move centre stage in the expanding Asian markets. But it comes with a vast baggage of new problems and tensions for which policymakers have been almost completely unprepared.

As a result the world will now have to turn, and turn with far greater vigour, speed and focussed determination than hitherto, to other energy sources to see what they can do to help in the approaching storms and difficulties.

CHAPTER SIX

THE OTHER FUTURE

෨ඏ ෨ඏ ෨ඏ

The other plentiful supply sources which we could be tapping NOW. How transport's mineral oil needs could be halved and halved again. Energy all around us, but that is not the problem.

෨ඏ ෨ඏ ෨ඏ

If energy policy experts and political leaders in the main oil-consuming countries could lift their eyes for a moment from worthy but distant carbon-reduction goals to the more immediate energy security dangers, they would see help coming rapidly along the road, with a little encouragement, not just from gas but from a whole range of other ready-to-deploy fuel energy sources, both fossil and non-fossil.

The problem with discussion of 'the alternatives' is that their proponents tend to be over the top in their enthusiasm and their detractors tend to be exaggeratedly negative. For the lay person, whether he or she be a lofty policymaker, a humble consumer or something in between, objective assessments are hard to come by. Nuclear power has its vocal supporters and critics. Coal is full of carbon, but clean coal could be a major factor – at a price. Biofuels sound more friendly but their costs, too, get deeply questioned. Wind farms split local communities, and are dismally inefficient, and so on. All of them spell headaches for governments as the rival lobbies line up with demands for tax breaks and subsidies for their preferred sectors.

What follows is at least an attempt at a consumer's and investor's guide. The approach is positive without being starry-eyed. Here first is the list of the front runners that could replace (at least partly) mineral oil:

1) The unconventional oils. These are the extra heavy oil and bitumen deposits now coming mainly from the vast tar sands of Canada. Shale oil also belongs in this group. The largest deposits, potentially containing as much oil as the whole Middle East, lie in the Green River area of Wyoming, Colorado and Utah. Unless the carbon is stripped out, these are bad news for the environment and low-carbon goals.

2) A variety of plant-based oils and oils from waste (biomass), including ethanol, bio-cellulosic ethanol and bio-diesel fuels, but their net energy gain may be small.

3) Enormous coal reserves in both China, the USA, the FSU (Former Soviet Union area) and in the UK which, with sulphur emissions purged out of them and carbon sequestrated and stored or used profitably elsewhere, constitute a major alternative resource to oil, but which energy policy has generally neglected to develop.

4) Windpower, clean but unreliable and environmentally intrusive unless localized and individualized as far as possible.

5) Solar power, where new technology is racing forward, even for less sunny climates, but costs are still too high.

6) And there is the biggest one of all – the nuclear power issue, which has had so much attention in the UK's so-called Energy Review and which falls into two parts. First, can existing nuclear power capacity round the world, which is considerable,[73] be preserved and its life prolonged? Second, can more plants be built, of what type and size and when?

Not included in the menu are a host of more specialized, and often localized, power sources like ground heat and tidal power, which can play their part, if only marginally.

Finally there is the prospect on the oil and gas production side of getting far more out of the present system and supply chain. (In the next chapter we will come to the biggest demand-side 'resource' of all, which means squeezing much more work, or output, much more efficiently, from the oil and other energy resources we consume.)

Long before oil reaches the consumption point, at every stage from wellhead to garage forecourt, or gas reservoir to household boiler and kitchen cooker, the system is riddled with cost-inflating inefficiencies. Refineries worldwide are not only inadequately suited to handling the heavy crude oils which increasingly make up supplies, especially from Saudi Arabia. Many of them – the European refineries in particular – are also very inefficient. North American refineries recover and distil about 90 per cent of the basic crude inputs they receive. In Europe the figure is 75 per cent. Middle distillates – the ones most in demand – such as diesel and jet fuel, are largely imported into western Europe, much of them from Russia. A more modern and efficient refinery structure, which again should have been built by now, would ease oil prices and supply pressures in Europe substantially and swiftly.

The Unconventionals

There is an oddity about the definition of the so-called unconventional oils. They are specifically excluded from measures of worldwide oil reserves. Being sticky, heavy and costly to extract, they were until recently considered outside the economic frame and not worth bothering about.

But as the costs of extraction and refining have fallen with new technology, while the cost of conventional crude has risen to $50 and higher, the heavy oils begin to look an attractive proposition.

If the quantities were small their exclusion from global oil reserve statistics would not matter. But they are not small.

They are huge. Some estimates suggest that the combined oil reserves of the Canadian tar sands and the Venezuelan Orinoco belt could be as much as two-thirds of total global oil deposits, thus making an even bigger nonsense of the 'running-out-of-oil' thesis discussed in Chapter Three.

The return to high oil prices has been a big blessing for countries such as Canada. The old joke amongst oil men used to be that whatever the crude oil price, oil from the tar sands of Athabasca, in Alberta, would always be $4 above it. Not any more. Oil extracted from the tar sands, by a long and costly process of squeezing out around one barrel of crude from over two tonnes of sand ballast, is competitive with conventional oil at anything above $50 a barrel. Investors and contractors are pouring into the state of Alberta and current output is running from all the mining areas at about one million barrels a day. It will rise to three million by 2015, or about 6 per cent of likely world oil needs (of one sort or another) by then.

The three biggest tar sands or oil sands areas of the world are Canada, California and Venezuela. But the tar sands of Alberta are much the largest (covering an area about the size of Ireland) and contain 175 billion barrels of oil. It is heavy and sticky – that is, it has a very high viscosity and flows badly – but it is there either on the surface or a short way down beneath and there is plenty of it. Only Saudi Arabia has larger reserves. Canada is now on track, from the tar sands sources alone, to becoming an energy giant – one of the world's leading oil producers and the third largest oil exporter. Oil substitute production (ethanol and other products) from corn and straw may put it almost at the top of the world energy league (see below on biofuels).

Jim Carter is the Chief Operating Officer of Syncrude, based at Fort McMurray in Alberta, and the largest oil sands operator in the world. A mining man by background who is constantly pressing for new technologies to lift and separate the huge loads of heavy sands, he believes that tar sands production is here to stay – for at least half a century ahead. This is because the economics of it all are at last beginning to look good and to look as if they will stay good.

The production cost of a tar sands barrel is now about $26. The process itself does consume a lot of energy (as well as a lot of water), and as the prices of gas and of diesel rise so does the cost of tar sands oil.

If safe, cheap and high-quality (light and sweet) oil were to continue to be fully available from the Middle-East producers, as in the past, there would be no chance of competition from the tar sands.

But that is not what is going to happen. Even with curbed growth of world oil demand, conventional mineral oil is going to stay expensive or dangerous, or both, to extract. Tar sands oil from Alberta at $26 looks a good bet. The world's investors and big energy companies agree. New technologies for getting it out, and new ideas for transporting and processing the colossal volumes of tar sands, are being developed all the time.

There are snags. Because it is so sticky, tar sands oil needs to be mixed with much lighter oils to get it to flow along pipelines. All this adds to production costs and marketing delays.

The same problems apply to Venezuelan heavy oil. The technique used there is to mix it with 30 per cent water and an injection of detergent. Most of Venezuela's heavy oil exports are moved to tankers and piped direct this way. But political instability in Venezuela casts a long shadow. If the country could settle down, if technical problems of upgrading and handling the sticky substance could be resolved, if... then tar sands oil and heavy oils generally would make significant inroads into oil markets soon. So far the contribution from both Canada and Venezuela is running at about 1.5 million barrels a day, less than 2 per cent of the world's total daily 'burn'. So it is early days. But the potential is vast.

Shale Oil ought to be on the 'immediate' list of unconventionals as well, since the deposits are large, mostly in politically reliable areas and already mined and processed in small quantities.

A tonne of shale oil rock can produce a barrel of crude oil. The biggest locations for shale are the Green River basin where the three US states of Colorado, Utah and Wyoming meet and, oddly enough, the northern coast of Estonia. The Estonians continue to rely heavily on oil from shale.

Why has it not taken off as an industry now that conventional oil is soaring in price and is generally going out of favour?

116

Unlimited sun from the desert – is this the next big thing? The technology is said to be 'about where mobile phones were ten years ago'.

The answer lies in the very heavy costs of extracting the oil from the rock (usually by 'cooking' it with steam) and in the extreme stickiness of the product, which makes pipeline transportation almost impossible. Also, enormous quantities of water are needed for the conversion.

But technology is pushing ahead and as the price of natural gas soars, trailing the price of crude oil, the shale possibilities begin to open out. Market trends and investor assessments will determine how soon shale oil joins the portfolio of new energy sources. It is probably best left that way.

How Much Oil Can We Just Grow?

Even without the tar sands Canada would anyway be lining up as a major energy power in the new pattern of things. The

117

reason lies in one word: 'Ethanol'. Hopelessly late in the day, policymakers and governments in the so-called advanced countries have started turning to energy possibilities which private enterprise and some less advanced countries (still patronizingly so mis-labelled) have long since seized and carried forward.

Plant-derived oils cannot obviously be an instant replacement for mineral oil – there are no such wand-waving magic solutions. But they can certainly be one powerful element in the portfolio of renewable sources of energy, via which big profits can be made and both big and small needs met.

Biofuels – ethanol from corn, grain, sugarcane, farm waste and even just from straw, and biodiesel from soybeans – are sustainable (i.e. they can go on and on being renewed), efficient (or can be), and have the potential to create jobs and economic growth in developing countries, reduce demand for costly oil imports, and address environmental problems ranging from desertification to climate change.

Campaigners for a lower carbon future ought to be cheering on these alternatives. But somehow the cheers are muted. The USA, constantly castigated by climate change 'experts' for lagging behind in energy matters and for failing to sign up to climate change commitments as required by the Kyoto protocol, is in fact one of the leading countries in moving to large-scale biomass fuel supplies. The infrastructure throughout North America necessary for inserting biomass firmly into the fuel supply chain is fast taking shape. The change is coming from the policies of individual states rather than from the federal government – further confirmation that the quickest and best 'solutions' to the energy crisis are going to be driven at the grass roots, that is by local initiative and enlightened self-interest, rather than by established (and remote) bureaucracies. Six American states already require all forecourt gallons to have a 10 per cent biofuel content. These include California, which by itself is the twelfth largest oil-consuming entity in the world and the fifth largest 'state' in terms of GNP.

In Canada a number of producer groups and private consortiums have been interested in establishing a large-scale fuel ethanol plant. Many of these groups have secured private investment funds for the construction and operation of the plant. They have also asked federal and provincial governments for tax relief for fuel ethanol production. The federal government is interested in fuel ethanol for three reasons.

First, a fuel ethanol sector would provide an additional domestic market for corn production in Canada, keeping the farmers happy, never an easy task. For Agriculture and Agri-Food Canada, the development of this market would decrease the agricultural sector's dependence on volatile international markets, heavily influenced by subsidies, production surpluses and protectionist policies all round the world, especially in the European Union.

The second reason for government support is the recognition that ethanol really does score environmentally. Despite the debate about its net effect on carbon emissions, when it comes to aromatic emissions and other poisonous outflows ethanol is definitely ahead. Claims that ordinary gasoline has killer side effects, including causing cancer, are in the 'nobody really knows' category, although whole books have been written asserting just that. But ethanol-blended gasoline has been recognized by Environment Canada as easing the burden on the environment and has been designated an 'Environmental Choice' product. Finally, the development of this sector could have benefits in terms of employment, infrastructure and Canada's trade balance, and contribute to energy security. So it looks attractive all round.

The Less Good News

But there are a few bits of bad news about biofuels. First, using agricultural products as a major energy source sounds good but tangles the issue up with the traditional political minefield of farm subsidies generally. American congressmen, and politicians everywhere in the industrialized world, get a gleam in their eyes when they see the possibility of huge new markets opening up for their heavily protected farmers. Naturally they want the existing subsidies and supports – of the kind which have paralysed world trade liberalization and led to the breakdown of trade talks at Doha in 2006 – to stay and even be increased. So the whole process could become interwoven with farm support lobbying, severely distorting the true economics of biofuels and upsetting food supply patterns as well.

Farm politics explain why the USA keeps a 54 cent per gallon tariff on imported ethanol – to keep out cheaper Brazilian supplies and protect US agriculture.

Second, environmentalists are deeply opposed to the destruction of rainforests to make space for planting soya crops ands sugarcane. If that is the price of independence from fossil fuels, is it worth it?

It is not at all clear whether there is any useful net energy gain by growing plants and transforming them into ethanol. This is just the sort of issue where energy security pulls in one direction and concerns about carbon emissions pulls in another. It explains, too, why low-carbon campaigners are doubtful about biofuels.

Back in the USA, the Department of Agriculture (USDA) in 1995 concluded that we got back about 25 per cent more energy in the ethanol than we invested in growing corn and transforming it. On the other hand, Cornell's Prof. Pimentel concluded in 2004 that we ended up with 29 per cent less energy. The issue seems to be a giant accounting problem, revolving around the definition of the boundaries of the system: how can proper account and measurement be established of the embedded energy in supplies and equipment used, and of the energy embedded in co-products.

If the USDA is correct, there is some net energy gain in making ethanol, but would it make economic sense (geopolitical sense is another matter)? Five gallons of ethanol have to be produced for every gallon available as transportation fuel. The other four gallons would be consumed within the ethanol production infrastructure. That would obviously have a tremendous impact on the land area required for ethanol to replace current gasoline use.

There are other complexities, too. The use of ethanol is based on the idea of taking an agricultural waste product people currently pay to get rid of and convert it into fuel. In reality, agricultural waste product is seldom wasted. It is often returned to the soil to maintain its texture and fertility, which is critical to agriculture and therefore to the survival of human civilization. If that waste product was instead diverted to ethanol production, there would be long-term consequences for agricultural productivity. Or to put it in more homely terms, if the garden compost is burned for fuel, something has to replace the compost.

Car and truck makers would need to build more slightly adapted cars, as has already happened in Brazil (see p. 124). The cost of adding this capability to new cars has been esti-mated at roughly $100 a vehicle. And ethanol would need to be much more readily available at gas stations.

The production of energy from biomass involves a range of complicated-sounding but quite advanced technologies. These include – wait for it – solid combustion, gasification and fer-mentation, among others. These methods all produce liquid and gas fuels from a diverse set of biological resources – tradi-tional crops (sugarcane, corn, oilseeds), crop residues and waste (corn stover, wheat straw, rice hulls, cotton waste), energy dedicated crops (grasses and trees), dung, slurry and the organic component of urban waste. The results are bio-energy products that provide multiple energy services: cooking fuel, heat, electricity and transportation fuels. It is this very diversity that holds the potential – and it could be within short-term reach – for the environment, for social and economic develop-ment, for energy security and for the switched-on and enterprising businessmen and women who invest and produce the new systems and products required to make it all work.

Biomass Old and New

There is nothing new under the sun. Biomass was the world's primary source of energy until the late 1920s. Today about 10 per cent of the world's energy use is still derived from biomass; however, this average masks the far greater importance of bio-energy in less developed countries in Africa, Asia, and Latin America, where its share is as high as 80 per cent.[74] The potential contribution of modern biomass energy services to a new energy paradigm could be really big.

Many regions and countries, including the European Union and Argentina, have adopted national targets of 5–10 per cent displacement of mineral oil with plant-based biofuels. Shell is voluntarily applying a 5 per cent biofuel content quota in Germany and the Renewable Transport Fuel Obligation in the UK requires the refiners to add 5 per cent to every gallon sold. But the policy is a furtive one and nothing tells the motorist at the pumps what they are buying or why they should demand more of it.

In booming Asia, Thailand recently implemented a biofuels programme that includes tax incentives and low interest loans for processors, to help reduce its dependence on imported oil and create a new market for its high tapioca yields. Malaysia is developing a palm oil supply chain, from palm groves to refined car fuel and power stations. Incentives for the production of biofuels are being put in place around the globe.

The potential for bio-energy to reduce global greenhouse gas emissions varies, depending not only on the feedstock conversion technology but also on the methods used to produce the feedstock. For example, ethanol produced in industrialized countries from corn may reduce life-cycle greenhouse gas emissions by only 10-30 per cent compared to oil, whereas ethanol produced from sugarcane or cellulose may reduce it by 90 per cent or more.[75] In both cases, greenhouse gas reductions increase dramatically if agricultural practices are adopted that absorb carbon dioxide back into the soil or drink it up in other ways (lots and lots of trees and tall grasses), and are less intensive in their use of petroleum-based fertilizers and fuels.

The transport sector, which uses half the world's conventional oil supplies, could become an early winner from biomass developments. On current trends, global oil use and carbon dioxide emissions in the transport sector will nearly double between 2000 and 2030.[76] Increased use of bio-energy fuels such as ethanol and biodiesel could help change this picture by offering an important low-carbon alternative to petroleum long before this. Production of biofuels, especially ethanol from grain and sugar crops, has been increasing dramatically in recent years, and some countries are already getting their biofuel act together and using these fuels in a big way in the transport sector. These do not yet include most of Europe or Asia, but the much criticized Americans are moving ahead remarkably fast, and the Brazilians have already got there.

So far, the most popular path for bio--energy use in the transportation field has been the conversion of traditional crops, like sugarcane and corn, into ethanol, either to be blended or directly used in internal combustion engines. Soybeans, jatropha and other oilseed crops can also be converted to biodiesel fuel and used to extend or substitute fossil-derived diesel fuel.

There is a third variety of ethanol called cellulosic ethanol, which comes from corn cobs or even weeds and general agricultural waste. Although biofuels offer a whole range of promising alternatives, ethanol constitutes 99 per cent of all biofuels in the United States. Much of the analysis and public debate about ethanol has focused on whether ethanol is a plus or minus in net energy terms: whether manufacturing ethanol takes more non-renewable energy than the resulting fuel provides.

Big money and big names have put enormous sums into ethanol. Vinod Khosla, a founder of Sun Microsystems and a partner at Kleiner Perkins Caufield & Byers, the Silicon Valley venture capital firm, has already invested tens of millions of dollars in private companies that are developing methods to produce ethanol using plant sources other than corn. Mr Khosla isn't the only big-name entrepreneur to embrace ethanol. The ebullient Richard Branson, chairman of the Virgin Group, plans to invest $300–$400 million to produce and market ethanol made from corn and other sources. Bill Gates has also made a move into the ethanol market, with an $84 million investment.

So despite all the doubts and queries the case for ethanol, and for biomass generally, is getting powerful. Soaring energy prices have made corn-based ethanol more competitive, while research advances in breaking down cellulose into simple sugars have cut the cost of making ethanol from other sources.

A very little way ahead, types of ethanol could become genuinely cheaper to produce, unsubsidized, than gasoline. Even if petroleum drops to $35 a barrel, innovating technology could still put ethanol very much on the map.

The Brazilian Experience

A 'shock' for conventional thinking about low-carbon goals and green energy ambitions is that one major nation has already almost arrived, via biomass, at the post-oil age. Brazil has rebuilt its economy on ethanol and other biofuels, using sugarcane as its main energy source. Conventional oil imports are closing down and exports of Brazil's 1.8 million barrels a day of conventional crude oil to other less 'advanced' countries increasing. Is 'less developed country' the right phrase to discuss a nation advancing so successfully to new technologies? As in so many other fields the old categorizations no longer fit the new international pattern rapidly emerging and both shaping and shaped by the global energy crisis.

Brazil has proved that ethanol can be made competitively from sugar. The cost of producing ethanol from sugar – including raw materials and processing – comes out at $6 to $7 per gigajoule versus $14 a gigajoule for gasoline.[77] But cellulosic ethanol, the kind produced from non-food plant matter, has some advantages over food-based ethanol. Because cellulosic ethanol is derived from plant waste, wood chips or wild grasses, it would not require costly cultivation; that would mean savings on labour, pesticides, fertilizers and irrigation (and probably farm subsidies as well). It is also somewhat superior to corn-derived ethanol in reducing greenhouse gas emissions, so it looks a better deal all round.

Today, the nine vehicle makers in Brazil (including General Motors Corp. and Ford Motor Co.) make 'flex fuel' cars that can burn gasoline, ethanol or any mixture of the two. With 'flex fuel' cars you get lower mileage per gallon, but the cost is so much less than gasoline. Ethanol sales in Brazil reached nearly five billion gallons in 2005, which would displace 323,500 barrels a day of gasoline.

Pipelines and storage capacity at ports and refineries are the weak links in Brazil's dreams of massive ethanol exports. To fix that, foreign investors are expected to pour $3.6 billion into the business over the next five years. Brazil's largest foreign ethanol customer is the United States, which bought 90 million gallons last year. This was despite the protectionist barrier maintained on all ethanol imports to cocoon U.S. producers. Congress gives oil refiners and blenders a tax break of 5.1 cents a gallon for gasoline mixed with ethanol. But it hasn't been enough of an incentive for Gulf Coast refiners to bring ethanol to Texas, even though the biofuel sells for far less than gasoline.

Not all ethanol is the same. It is made from sugar in Brazil and chiefly from corn in the United States. In Brazil, ethanol is sold as a stand-alone fuel and also gets blended into gasoline at 24 per cent concentrations. Half the new cars sold in Brazil now are 'flex fuel' vehicles that can burn either gasoline or ethanol. Lower production costs mean pump prices for ethanol are more than $1 a gallon less than gasoline.

In the USA, by contrast, ethanol is blended into gasoline at 10 per cent concentrations and can be burned in regular gasoline engines. A small amount of ethanol is sold in concentrations of 85 per cent for 'flex fuel' cars that can burn either gasoline or ethanol. In some markets, consumers get discounts of as much as 10 cents a gallon for gasoline with ethanol. Ethanol isn't blended into gas in Texas.

Since the 1970s, Brazil has saved almost $50 billion in imported oil costs – nearly ten times the national investment through subsidies – while creating more than one million rural jobs. There were wobbly moments for Brazil, when oil prices dropped sharply in the 1980s and again in the 1990s. But the Brazilians somehow stuck to their commitment, and now it is clearly paying off.

Brazil's experience shows how government leadership and smart policies can reduce dependence on imported oil while boosting local economies. It's a success story that a growing number of US political and industry leaders are eager to emulate. But if Brazil could ramp up its use of ethanol and diminish its dependence on foreign oil, can other countries do the same?

Can or Should the USA or Europe (or anywhere else) Follow Brazil's Example?

US ethanol production is also climbing fast and will consume about 15 per cent of the corn harvest in 2007. The development of the ethanol industry has been driven at least as much by agriculture as by concerns about oil. That's been successful in getting the USA to a consumption level of a little under four billion gallons a year. But if the country wants to be serious about replacing large amounts of imported oil, it has to be thinking ten times that amount.

The fast-growing US biofuels industry calls Brazil a model for what could be accomplished. But what works for Brazil won't necessarily work in the United States. The Brazilian government has for decades subsidized an ethanol delivery system that puts pumps in every gas station. And Brazil has a much easier path to energy independence. It imports a mere 240,000 barrels a day, or just over 10 per cent of its oil. By contrast the US imports 13.7 million barrels a day, or nearly two-thirds of its needs. Brazilians claim their sugarcane ethanol is made for one-half to two-thirds the cost of US ethanol made from corn. One reason is that Brazil has a better climate for biofuels. After the 8-month, frost-free growing season in Brazil, sugarcane yields high volumes of the new fuel.

Corn and other crops grown between winter freezes in the US don't have the same yields per acre of crop. Brazilian sugar mills use the cane husks as a boiler fuel, and they send surplus electricity into the national grid. The ethanol plants popping up all over the US corn belt don't yield as much net energy because they are consumers of natural gas, coal and electricity. And US farmers don't have Brazil's cheap land and labour costs. Mario Gandini, manager of the Sao Martinho sugar mill and distillery, said his cost of producing ethanol is $200 per cubic metre – which works out at 75 cents a gallon. 'We know no country can beat us in production costs', he said.

While 'flex fuel' sales are booming in Brazil – and in April 2005 they accounted for nearly half the country's new-car sales as high gasoline prices lured consumers back to ethanol – in US markets they have not yet made a big impact, mainly because of the limited availability of high-ethanol-content gas blends. John Felmy, chief economist with the American Petroleum Institute, says prices for ethanol may be lower – but so is its energy content. A car travels about 30 per cent farther on a tank of gasoline than on ethanol. And Texas gasoline suppliers use MTBE, a petroleum-based fuel additive, to achieve the cleaner-burning characteristics that ethanol delivers for Midwestern markets. 'You just can't switch overnight from MTBE to ethanol. You have to have a blending facility, rail cars and other types of transport to get it where you need it', Mr Felmy said.

Country	Amount (million litres)	Share of World Production (per cent)	Primary Feedstocks
Brazil	15,110	37	Sugarcane
United States	13,390	33	Corn
China	3,650	9	Corn, cassava and other grains
India	1,750	4	Sugarcane, cassava
France	830	2	Sugar beets, wheat

SOURCE: See *State of the World 2006*, Chapter 4, Endnote 13.

So Yes to Biofuels Generally?

The notion of a new energy paradigm conjures images of automobiles propelled by silent and super-clean hydrogen-powered engines and solar panels illuminating houses and streets. That is possible but it could be 50 years away. Yet we are, say, five years – possibly much less – from potentially serious disruption in world energy security – for which feeble and inadequate preparation has been made and every opportunity for remedial action ignored by sleepy policymakers who still cling to the belief that cyclicality plus a few good speeches and reviews about the long-term possibilities will bail us out. The real and much more urgent question is what can be done now.

The answer is not everything but definitely a good deal. The diversity of feedstocks that can be transformed into useful energy means that almost every country is free to develop its own unique domestic energy industry. Combined with enhanced production practices in the agricultural and forest sector, this would also help the viability of rural economies and reduce the exodus of rural populations to urban areas. Each country can shape its own strategy, so can each region, so, for residential and small-scale users, can every community.

In sum, ethanol is a promising part of the energy mix that could come-on-stream quite fast. It is clean burning. Costs are falling fast and market forces will do their work. It may not quite yet look competitive with oil, but the full cost of conventional oil dependence, in terms of foreign policy and in terms of geopolitical dangers, present and future, has to be taken into account. When it is, the sums begin to look quite different and the case for plant-based oil products suddenly becomes much stronger.

Meanwhile, the whole prospect could obviously be a blessing for farmers, if, but only if, the politics are wisely managed. The stuff can be sold at the existing garage pumps without much modification. Like hybrid cars, it is all going to happen faster than the 'experts' or policymakers predicted or expected.

The big oil and petrol marketing companies are stirring. Shell already includes 10 per cent of biofuel in its standard gallon of unleaded petrol and its standard gallon of diesel. This practice could be made mandatory or better still a mixed gallon could be taxed at a lower rate. At some point as the oil price climbs in the near term (and even if, as is likely, it drops temporarily and then shoots up again) it will also become cheaper for the oil companies to deliver. As always with the launch of new products capital investment gambles will have to be taken and the difficult initial start-up period navigated. But combine this with spreading public demand for the greener fuel and the pattern of demand for conventional petrol could be sliced down dramatically and in very short order.

In addition, there is the simple fact of convenience at individual household level. Economists often overlook the hassle factor in evaluating alternatives. For a householder to be told that this or that new technology will pay him or her back in saved fuel bills in x number of years is of little interest if it

Canada's tar sands – a vast new oil resource. From a safe and reliable country, but hardly an environmental blessing.

means large initial capital outlays and having to tear out and replace existing cabling, fuse boxes and the like, or having to re-roof the house with solar tiles.

For the household which uses heating oil much the best 'solution' in the near-term is simply to see the oil tank behind the garage filled with vegetable-based kerosene instead of mineral-based – and hopefully at a much lower cost. That is the attraction which biofuels hold out – a smooth and cheap transition. When the word begins to circulate that this can truly be achieved a tipping point will indeed have been reached and an immensely swift transition will take place. Biofuels could, given the right encouragement, supply 25 per cent of America's or Europe's oil needs within the next five years.[78] But that would need not just market action but policies and attitudes which are just not in place – yet.

Overall, bio-energy has to be viewed not as a replacement for oil, but as a large potential element in rebalancing the portfolio of renewable sources of energy. As a very informative House of Lords report emphasized,[79] if energy security is the name of the game then it makes sense to help biofuels along briskly, and even to encourage imports (from that Brazilian sugarcane again).

But if cutting carbon emissions is at the top of the agenda, as it is in the case of the UK Government, then the scene becomes much more complicated and the hesitations creep in. Some bio-fuels probably do emit more greenhouse gases by their production than they save when consumed. Some form of carbon certification then has to be introduced, raising the possibility of more bureaucracy and administrative complexity. Suddenly the shine begins to go off the plant-based alternatives.

All this underlines the distinction between putting energy security first and putting carbon reduction first. That is the difference between practical and serious measures and green dreams. Forget about security in the shorter term and the longer-term goals may never be reached.

Coal – the Forgotten Jewel or the Messy Past?

Enormous reserves of coal lie under the earth, much of it, although not all, in politically fortunate and reassuring places. North America and the UK have very extensive reserves – enough to meet energy requirements for hundreds of years. The USA is truly the 'Saudi Arabia' of coal, while the UK could well claim to be the 'Kuwait'. China, India and Russia, while far from being paradises of stability, also have massive coal deposits. Together, America, China and India have about two-thirds of the world's coal reserves.

Furthermore, it is the least expensive unit of energy, at least when burnt conventionally, and it is the source base for a range of products equal to, or exceeding those from oil, including high-grade gasoline,[80] the cleanest kind of diesel (in the form of dimethyl ether), numerous chemicals, hydrogen and many other gases.

But coal has a truly rotten image. It is associated with grim underground dangers and accidents, with belching pollution, Dickensian misery, grime, strikes and political unrest. It may have powered the Industrial Revolution, but its story is one of stupid and greedy coal owners and even more stupid labour leaders who, like Arthur Scargill in the UK, have led their mining industries almost to obliteration. Worse still in today's context, coal is full of carbon and full of sulphur. Who wants to go back to coal?

Yet the question is a misguided one because it ignores the present situation and it misses the point that new technology could change everything – and may already be doing so; 60 per cent of US electricity comes from coal-fired power stations and 120 more coal-fired stations are being planned or under construction in America alone. In the UK the figure is just under 40 per cent most days, although when gas runs short or the pressure goes down, the coal contribution can rise to as much as 80 per cent. China, while trying every avenue for increased energy production, is heavily committed both to many more coal-fired stations and to extensive gasification of coal. Texaco has already sold China eight coal gasification plants and more are to come.

Coal can also power transport. The famous Fischer-Tropf process, used during the Second World War to fuel the Nazi War machine, can turn coal into diesel and into benzene of high quality. South Africa under the apartheid regime took up the same process to supply its petrol needs – again not a very happy parentage. But it has been estimated that if US coal reserves were to be processed this way, this would produce enough power for the needs of the USA for at least 250 years. No running out there!

So the question should be rephrased to ask how this still enormous coal burn round the planet can be matched with the demands for cleaner air, less pollution and fewer carbon emissions, as well as with better and less risky mining methods, commensurate with modern standards, and, of course with greater energy security. If these standards and goals can be met, and the total production and processing costs at the end of it all still come in well under the present or prospective oil price (with all the risks, collateral and hidden costs of oil dependence factored in), it is hard to see why coal should not again become king.

We are back here with problems of fashion and vision. If all eyes are on reducing carbon emissions in the long term, then the misplaced instinct is to shy away from coal altogether and neglect coal's possibilities. The technologies for cleaner coal-burning, and 'sequestrating' the heavy carbon emissions, are there – although as yet unproven. But what does not get tried does not get costed, and the general antipathy to coal 'solutions' persists. The research and development needed to deal with coal's dirty past stays on the low priority list. That is just what has been happening. In the UK, despite endless and patient prodding and questioning in Parliament at Westminster,[81] the required innovation-forcing research in the coal sector has been woefully slow and limited.

Sulphur emissions, which cause immediate pollution and hang over many Chinese cities in an appalling black and yellow cloud – as they once did over London, creating the famous killer smogs of previous eras – can be eliminated completely. Carbon from coal-burning which has been 'sequestrated' can then be piped into undersea caverns, or used by offshore oil operators to increase underground pressure in oil-fields and extract more oil.

Through another technology, called Integrated Gasification Combined Cycle, which unlike sequestration has already been developed and tested, coal can be cleaned of most impurities and turned into synthetic gas for use in turbines – a far cry from the old days of coal gas. Small amounts can be, and are, sold to horticulturalists as a highly effective growth stimulant to flowers and vegetables.

In the UK in particular the issue is all bound up with national psychology and history. Until the 1970s the UK relied very heavily on indigenous coal for its electricity. But the miners' strikes of that decade and the early 1980s led to the obvious conclusion that this was one self-sufficient fuel which could not be relied upon. Hence the deep antipathy to going back to coal, even though conditions have changed totally 20 years later.

The mining communities in the UK contained a mixture of the very best qualities to be found in the British people, together with ripples of fear, apprehension and insecurity. Fear makes a bad counsellor and in the British case led to the acceptance of bad and ill-motivated leaders like Arthur Scargill and Mick McGahey, who misjudged everything and led the miners into eventually disastrous militancy. They misjudged their own power to bring the country to a standstill, having very nearly succeeded in the early 1970s, and egged on by the cognoscenti they misjudged the determination and skill of the British Government and people not to be held to ransom by miners' leaders ever again.

As a result they led the industry to destruction, backed by a political left who cheered them on, and coal became almost a damned commodity in Britain. All those groups and personalities on the left and centre-left, and all those publications, who 'backed' the miners' strike of 1983–84, and who encased the 'cause' in sentimentality, carry the responsibility for inflicting destruction and misery on a whole industry and its communities, barring the way for the re-enthronement of a modern domestic coal industry that should have begun to take shape long before now, but hasn't – a good example of how left-wing 'compassion' can be really cruel.

Even so, moves could begin now, if policymakers were not still asleep, to bring coal back to the centre of energy planning. An obvious sequence would be: a) move much more electric generating capacity back to clean coal and away from the heavy gas burn which has now developed (in the UK, 39 per cent of all generation); b) use natural gas – much cleaner than oil – for conversion to liquid fuels for powering all cars and trucks; and c) use the remaining oil, at its continuing high price, for aviation and for some petrochemical and materials processes for which no substitute could be found.

All that could happen now with a little encouragement from public policy. Nothing of the kind got a mention in the UK's much trumpeted Energy Review Conclusions,[82] which were far more concerned with carbon reduction (40 years on) and new nuclear plants (ten years on).

Coming with the Wind

Electricity from wind pylons is a source which has definitely not been left to market forces. The politicians and lobbyists have entered the field with a vengeance. Subsidies and tax breaks abound and this may all prove to be the industry's undoing (of which there are already signs).

No energy-supply system can run at all-out capacity and nor should it. There has to be a margin of spare. Coal-fired power stations usually run at about 75 per cent capacity, nuclear plants (when they are operating correctly) as high as 92 per cent. But wind farms are doing well to get to 35 per cent – for the obvious reason that the wind blows intermittently.

This has not stopped both European and North American authorities getting excited by wind power and putting quite a lot of taxpayers' money into them. The EU has given itself the target of producing 20 per cent of energy needs from renewables by 2020 and the UK has given itself an additional target of 20 per cent by 2015.

Whereas mountainous countries like Norway or Switzerland have no difficulty in meeting this sort of target via hydroelectric schemes, the UK has left itself little choice but to reach for these goals via wind power. Germany and Denmark have also invested heavily in wind power.

The snag is that a lot of wind power is needed, with a lot of conventional power station back-up, to generate the steady currents needed by industrial consumers, by urban communities and by homes and offices. By 'a lot' is meant erecting about 1,500 large wind pylons to produce the same amount of electricity as one nuclear plant (say 1,000 megawatts).

Locating these forests of monsters on land (they are often twice the height of Nelson's Column in London's Trafalgar Square) is bound to intrude on the landscape and upset nearby homeowners and nature-lovers – and certainly does. Add in the roads to get supplies and service trucks in and out, and the pylons to export the electricity, and the sub-stations to transform it from Direct Current (which comes from wind turbines) to Alternating Current to feed into the Grid, and a recipe for full-scale confrontation emerges between environmentalists and pylon builders (and between environmentalists in love with 'green', renewable energy, and those in love with beautiful downland, silent marshes and unsullied skylines).

A small, self-assembled, wind pylon at the end of every garden or orchard (for those lucky enough to have one), with wiring to the home which is not too complicated, seems a real possibility, although it has yet to be developed. But massive wind farms to replace base-load needs for daily electricity, in a world which is set to become increasingly electricity-based, seems a fantasy. The recourse of putting wind farms out to sea looks good, and attracts politicians' posturing, but leaves a host of questions unanswered about corrosion, shipping safety and bird life, as well as shoreline intrusion where the current has to be brought on land.

At best, wind power can add a few percentage points to electricity supply at something near a competitive price (although the economics are heavily confused by subsidies and levies on other energy sources). At worst, wind power can turn out to be extremely unreliable, and some countries (e.g. Denmark) have wound down their programmes of wind farm building, while major corporations (e.g. the German giant E.On) have found wind power hopeless, unreliable and inadequate as a power source. There is also the unavoidable problem that fossil fuel power stations have to be standing by when the wind drops, emitting carbon heavily as they are switched on and off to meet the wind's vagaries.

Wind power may have a small place in the long-term energy future, mostly at the household and residential level. But its impact on the near-term energy security scene will be minimal and the enthusiasts (and lobbyists enriched by subsidies) who have rushed into extensive wind farm developments will be seen in due course to have taken public opinion for a colossal ride, although this may take some years to emerge.

Power from the Sun

Is the desert a solar powerhouse? Perhaps. The means certainly exists to collect solar power in very large volumes through the so-called Concentrated Solar Power Technology. This can produce heat at up to 300°C, which is then used to drive steam turbines and generate electricity. If these plants are located on the seashore they can use sea water for cooling and then – and this is a nice addition – turn it into fresh water, the other key resource which hotter regions obviously need.

It sounds good. The central issue is, as usual, cost. At present the expert technicians in this field claim that solar power can compete with oil at $50 a barrel. That does not give much of a margin for investors and businesses looking for a big return from quite big risks.

The sums could of course change, and probably will. It is conceivable that solar power could yet be 'the next big thing'. The technology has been described as 'about where mobile phones were ten years ago' – admittedly by a big Chinese investor in solar devices. But he may have a point.

Back in Europe, Germany has gone solar. Or to be more precise, the German Government has introduced subsidies and tax incentives for installing solar panels and tiles in homes, and building larger-scale solar panel complexes for industry, which has led to a remarkable surge in demand for solar equipment. The overall figures are still modest but the increase from almost nothing to a serious chunk of the power system is notable. The European Union currently has some 1,500 megawatts of solar power installed producing electricity for heat and light (about the size of one large nuclear power station); 65 per cent of this is in Germany, and that amounts to about 1.5 per cent of Germany's total electricity consumption. Not much, but a useful contribution which could grow in the immediate future.

Technology is producing more and more-sensitive and efficient coatings for collecting the sun's heat and light. As always the real barriers are not so much the general enthusiasm, which is considerable, but the actual capital and disturbance costs of installing panels in the home. Solar-sensitive tiles, looking very nearly like ordinary roof tiles, can be installed, ideally when a house is being re-roofed. But that does not happen often, and the costs and hazards of workmen crunching about on the roof are proverbial.

Sales patter about saving on fuel bills in the years ahead impresses very few except the most gullible. The question is whether installation and connection with existing electric wiring is cheap and easy here and now. Sunshine is of course necessary as well, although the newest solar technology allows operation with surprisingly little hot sun. Light is nearly enough.

Nevertheless, as a result of technological advances that have increased the reliability and reduce the cost of solar power for electricity generation, as well as tax incentives for solar electricity generation, large solar projects are popping up around the world. In the USA, FPL Energy owns and operates nine plants in California's Mojave Desert that comprise the Solar Energy Generating Systems (SEGS). Built between 1986 and 2005, the SEGS facilities produce up to 354 MW of electricity for the power grid that supplies the Los Angeles area. Stirling Energy Systems has revealed plans for two large-scale facilities that would comprise what has been called the largest solar power project in the world.

The company has proposed to build a 500-MW solar farm in the Mojave Desert and a 300-MW facility in California's Imperial Valley. In a process that is said to be two to three times more efficient than conventional photovoltaic cells, the plants will use 40-foot tall curved dishes that concentrate the sun's energy on an engine filled with hydrogen; as the hydrogen is heated, it expands and drives the engine's pistons. Construction on the facilities is expected to begin in 2008.

In both Germany and Japan giant solar panel facilities have been built in an attempt to generate big volumes of electricity for industry. But doubts remain about costs. At a certain stage large-scale solar costs will come down as oil prices go up and the two lines will cross. But when? If policymakers had asked that question five years ago we might now have the answer. As it is solar power may just about be in time to help surmount the coming supply issues. But the contribution could have been much greater much sooner.

The Future is Nuclear – or is it?

More nuclear power stations may in due course be built, replacing the considerable number round the world now in operation and maybe expanding nuclear's overall share of energy supply. There is debate as to how much carbon is created over the full lifetime cycle of a nuclear power station's existence. The opponents of nuclear power say it is high. But what no one can dispute is that when it comes to actual power generation, nuclear power is virtually carbon free.

So the first question is whether new nuclear plants can be built in either the USA or Europe in time to ease the energy pressures immediately ahead. France, which took the bold decision to build no less than 40 nuclear stations (of the pressurized water type) between 1965 and 1985, and draws over 70 per cent of its electricity from nuclear sources (58 plants in all), has obviously answered in the affirmative. It is even now involved in a programme of refurbishment and replacement as some of its plants wear out.

For other countries, like the UK, which have let their nuclear plants run down without replacement plans, it is now a matter of how quickly they can go into reverse and mobilize the resources to build again.

The timing does not look good. Instead of ensuring that replacements were ready to come forward, UK policy over the last decade has been to let its entire nuclear capacity (22 per cent of total electricity supply at its peak) be gradually phased out.

Changing direction now, as indicated by the UK Government's latest utterances (2006) on the issue, is not just a question of pressing a button. Teams and expertise have to be built up again. Decades of fashionable emphasis on green 'alternatives' have virtually eliminated nuclear engineering from the university career path, while a great deal of public fear, and little understanding, have been allowed to persist about the waste-handling issue.

There is also a wider unease that nuclear power stations have to be very large and therefore belong to an age of heavily centralized power generation more appropriate to the Stalinist period than the more decentralized microchip age. Big new power stations mean big new transmission systems. The UK nuclear industry is already talking about another £1.4 billion to be spent on upgrading the national grid to cope with new nuclear plants.

Should every home have one?

Government approval for new nuclear stations, plus portentous announcements in Parliament, do not mean that the plants will necessarily ever be built. In 1980 the Government of the day announced plans for 11 new large nuclear power stations, all to be based, so it was decided (after considerable and very prolonged debate), on Pressurised Water Reactor technology, as developed primarily by Westinghouse. Only one was ever built – the B plant at Sizewell on the coast of Suffolk, and that took 12 years to get into operation (eight of them in planning arguments).

This time the British Government want to see corners cut on the planning time taken, proposing that planning enquiries be confined to local environmental issues and not be allowed to drift into broader debate about the virtues or otherwise of civil nuclear power itself.

Finance will be a central problem. With all the doubts and delays, and the continuing political hostility, it becomes almost impossible to estimate what final costs will be and whether the electricity produced will be anything like competitive with rival energy sources. This makes new large stations, similar to Sizewell B, almost impossible to finance. A cat's cradle of levies, subsidies and surcharges has to be devised to ensure decent returns, in the end amounting to a heavy charge on Government and the taxpayer and/or electricity consumer (which is nearly everyone). So these are decisions that will have to be taken by Ministers and their civil servant advisers. The track record ensures that these people, however dedicated and sincere, will almost certainly get the key decisions wrong.

Behind the continuing public unease, which translates itself into more delays and still higher construction costs, lie persisting folk memories of things going badly wrong at Three Mile Island in 1979, when the core of one of the reactors melted down, and the Chernobyl disaster in 1986, when the reactor core actually exploded.

Worries also persist about waste storage. Two decades ago it was recognized that much the safest way of handling radioactive waste from spent fuel was to encase it in glass (vitrification) and bury it deep in the earth in stable rock formations. Endless enquiries, hearings, reports and pronouncements then ensued, all coming back to roughly the same answer. That is where the matter lies. More worries still surround the issue of decommissioning plants at the end of their operational lives, on which there are many guesstimates but few hard facts. The costs are horrendous and the US has already spent many billions of dollars in cleaning up from past programmes.

Behind all these doubts again is the nagging concern about the intertwining of civil nuclear power, nuclear weapons and terrorism. Unfortunately the route to the bomb, or one of the routes,[83] begins on the same track as the route to civil nuclear power. At a certain point the uranium needed to heat the core of a civil reactor can be enriched to 'weapons grade' and bomb-making comes into reach, although it takes time. More worrying still, capturing the neutrons from U235 fission can be used to manufacture plutonium from which, so it is said chillingly, it is much easier to build a bomb or warhead.

Once built, how can ownership be controlled? The Non-Proliferation Treaty regime, already under severe tensions, is supposed to constrain world nuclear activity. But terrorists are not deterred by regimes, nor, it seems, are rogue states like North Korea.

At the International Atomic Energy Authority (IAEA) head-quarters in Vienna minds are turning to possible better ways of accommodating a world of expanding civil nuclear power with effective stoppers on nuclear weaponry. Could some kind of genuinely independent and international nuclear fuel bank be devised to meet *bona fide* needs for enriched uranium round the world and to handle spent fuel, but prove a more effective check on nuclear weapons development than the past regime? The answers are vague – not least the answer as to the way decisions would be made about 'loaning out' nuclear fuel to countries building nuclear stations, and by whom. But with the old non-proliferation regime looking increasingly tattered it is clear that new thinking is required to combine a big growth of nuclear electricity with a safer world.

All this adds to the deep unease surrounding nuclear power expansion, although this is undoubtedly partly offset by the low-carbon qualities of nuclear electricity. It may also be that the latest nuclear power technology can lead to much lower waste outputs and much smaller and safer plants, getting away from the Stalinist flavour.[84] But if these are the possibilities, the nuclear experts and authorities have kept them largely to themselves. Many explanations and reassurances are needed if there is to be a nuclear electric future.

This all suggests that nuclear power will continue to have a part to play in the longer haul (i.e. existing nuclear stations will at least be replaced), but 'long' is the word. In the shorter-term future, in which severe energy disruptions loom, nuclear power will have little part to play. It will remain the stuff of Ministerial speeches and bar-room chatter about energy futures. But responsible political leaders and opinion formers should have their minds focused on bigger and more immediate issues.

Conclusion

Since energy is all around us and in unlimited supply, debates about the long-term supply of all the various resources and how to harness them are also virtually unlimited. But human society

141

lives in time and space and needs safe, reliable energy all the while, now, not in some remote and hopeful future. The possibilities examined in this chapter show that on the supply side the near-term problem of energy security can be tackled by vigorous and coherent actions. Drastically reduced oil dependency is well within reach, even though energy demand is going to grow massively.

The estimate for the USA is that over the next 18 years (to 2025) no less than 52 per cent of forecast oil use could be eliminated. And half of the 48 per cent remaining oil use could be met from biofuels. The other half (less than a quarter of forecast oil consumption by 2025) would also be met in part from unconventional oils, including tar sands, and from cleaner oils from natural gas. Applied to the UK the scene would be roughly similar, although the natural and frozen gas element might be bigger and the tar sands source (which after all is right next door to US markets) might be less significant.

Either way that would spell the end of oil dependence on the Middle East with all its endless feuds and threats, and it would be a reliable and realistic gateway and path to the greener, cleaner and more reliable energy future. Oil efficiency has doubled in America since 1975 and can double again. That is a practicable, near-term goal.

Handled right, the transition could not only be sustained but highly profitable, not least for the oil companies and energy industries themselves and the largest energy consumers such as the car and truck makers and the whole aviation industry, from plane-builders to airline operators (who are at present being crippled with rising fuel charges).

And that is just the supply side half of the story. The other half lies on the demand side – the far more efficient use of oil, the more efficient use of energy, the opportunities for using less energy – where, again, the opportunities are legion and the right measures, the right actions and the right attitudes can do wonders and save much grief. But neglect, inertia or the wrong measures and actions can leave the world in the disruptive chaos into which it is now fast slipping.

CHAPTER SEVEN

LESS IS MORE: WHERE TO TURN FAST AND FIRST

෨෨ ෨෨ ෨෨

How still soaring world demands for oil can be curbed, a
different pattern of energy consumption shaped, and
how a stronger lead given by both governments and
business can ease energy-related world tensions quickly
and enhance climate security prospects.

෨෨ ෨෨ ෨෨

The situation is unsustainable. Even though oil and gas deposits
exist deep beneath the earth's surface in huge quantities, there is
no possibility of continuing with the present rate of growth in
world oil consumption without running into further serious
economic and social disruption, with major political consequences, in the very near future.

The world may by now be burning up 88 million barrels of
oil a day, well over 1,000 barrels every second. We know that
in the calendar year 2006 there was an annual 'burn' of 31 billion barrels of oil, against which fresh reserves of some nine
billion – that is, mineral oil which can in due course be recovered and marketed – were found. The difference came out of
known reserves, mostly in the Middle East.

Of course there is a lot more oil around, some accessible at a cost, as in the Caspian Basin or offshore around Africa or still in the North Sea, some identified but so far unexplored in detail, some just guessed at. But the central, glaring point is that it all costs increasing amounts to extract. Even the oil in the 'easy' giant fields of Saudi Arabia is getting more costly to extract because these fields are more than half depleted and it is a simple fact of engineering and oil life that once fields are half depleted costs start rising fast.

Costs are anyway rising in these areas for other reasons. To meet demand the main OPEC producers have been pumping out oil at full tilt. The big spare tap that used to be so easy to turn on to raise production significantly at short notice is shut off, or if it is turned on only a trickle comes out. Any further surge in oil demand, or any sudden cut in supplies anywhere in the world in the present supply system, is immediately reflected in a shortage warning and a price blip as traders mark up their stocks.

The other reason costs are rising is that it is getting more dangerous to operate in the previously 'easy' regions. Danger spells risk and risk has to be paid for when investors put up their money. More capital up front, more spending on security, higher wages to persuade staff to work in personal danger – it all adds up to a bigger spend to get a barrel of oil out of the ground and moved to market, wherever the location.

All this implies that oil prices could stay high at least for the next few years – probably for the next decade. Oil being still a commodity, although one entwined with the politics at every turn, the price may swing wildly, with each drop being greeted by 'I told you so' cries and dismissal of warnings of dangers to come. But in practice any price relief will be temporary and seasonal. Oil will be forthcoming, alright, from many different regions, some of them quite new, but it will remain costly stuff.

The danger, already mentioned, is that any temporary price drop will be taken as an opportunity to give up on energy efficiency, to give up on alternatives and to give up on oil substitution. Pleading with the public about the longer-term implications of global warming will not be nearly enough to keep up the momentum, especially if exaggerations and inaccr-acies creep in. Memories are short. Reasons for tolerating any uncomfortable changes in daily life, and in matters as basic as power and other energy needs, fade very fast.

That is why in these pages the contention has been repeat-edly advanced that global energy transition needs a lot more than carbon controls. Short- and medium-term energy security issues need addressing just as urgently as long-term greenhouse gases.

There is no avoiding this dual approach. Substitutes and alternatives will also come along, but as has been shown in ear-lier chapters they will only appear at a price. Expensive oil has a knock-on effect throughout the entire energy system, quite aside from all the hidden costs incurred by governments (and their taxpayers) in attempts to safeguard world supply lines.

The only option for oil users everywhere is to make do with less oil. The ideal would be to use less energy generally, but in practice the world is set to consume much more – from one source or another. A less ambitious aim, but a fully attainable one, should be to use much less energy per unit of product or output, so that cutting out expensive oil does not just mean a switch in demand to other supply sources, thus forcing up their price, although to some extent it is bound to do so and is doing so already. For example, higher priced oil turns people not just to more economical oil use but to gas, which raises the gas price and has already put up the cost of electricity, and of heat-ing, lighting and almost every production process and service industry. To the extent that oil usage can be curbed further knock-on price effects will be eased, so this is where the imme-diate priority should lie.

It Has Been Done Before

It has been done before. Between 1977 and 1985 US oil con-sumption fell by 17 per cent, while GDP rose by 27 per cent.[85] The market and public demands showed the way, assisted by the iron policies of the Chairman of the Federal Reserve, Paul Volcker, who refused to accommodate higher oil prices with a cushion of easier money. Over the same period US net oil imports fell by 50 per cent and US net oil imports from the Arabian Gulf by 87 per cent, thus helping to weaken OPEC pricing power and paving the way for the oil price collapse of 1985–86 which was, if anything, too successful. That was the point at which the grandees of OPEC no doubt wished they had entered into a more cooperative dialogue with the consumer countries a few years earlier.[86]

Of course there is another possibility, although it can hardly be described as an option. That is to let oil prices zigzag on upwards and take the pain and disruption. The pain of trying to carry on with current consumption levels when oil lurched to, say, $200 a barrel would be very substantial. World growth would come to a halt, as it did momentarily during 1979–80, financial markets would be thrown into chaos, fuel poverty and hardship would increase greatly, whole industries would become paralysed (road transport, food and general retail distribution, airlines and travel), and developing country economies would spiral down. The international scramble for oil would become even more intense, with conflict replacing collaboration in many areas, such as Central Asia.

Middle-East turmoil would intensify as Western foreign policy floundered. Violence and terror would seize their moment as the whole energy system, operating under intense pressure, became more and more vulnerable to sabotage at key choke points and danger spots. In desperate attempts to curb energy costs the extra burdens imposed by low-carbon policies would be thrown aside and the additional disciplines demanded to cut carbon-emissions ignored. The international scramble for scarce oil would intensify and turn violent.

This book is about ways to escape this imminent route to disaster, so let us put this 'option' out of mind and concentrate, as policymakers everywhere should be concentrating, on the opportunities and possibilities for a smoother transition to lower oil dependence over the next few years. The goal is the modest one of how to make the ride less rough and more manageable. The right policies and understandings, combined with a willingness to learn from past mistakes, can ensure this.

The end of oil? At least for motor cars. President Lula da Silva of Brazil shows the way with a vehicle that goes just as well on alcohol (mostly from sugarcane) as on natural gas and conventional petrol.

We can begin on a highly positive note by recognizing that the opportunities offered by the transition are legion and could, if properly recognized, bring enormous benefits to richer and poorer societies, along with millions of new jobs, a new industrial pattern and a transformation of lifestyles on a scale comparable with the impact of the information revolution of recent decades.

The Two Revolutions

In fact the two revolutions – of information and of energy – play on each other and give each other momentum, spinning and weaving the geopolitical scene into a far stronger and more enduring tapestry than the miserable, torn picture confronting us today.[87]

This will be for two sets of reasons. First, because in a world less panicked by oil instability and less terrified that the

147

chief oil sources could be cut off or destroyed a more balanced and calmer foreign policy will have space to evolve. The heavy commitment of military resources, weaponry and manpower to Western intervention in the Middle East, with all its consequences and repercussions, would be wound down.

Second, because even with existing technologies, let alone the innovations just round the corner, business enterprise needs only a small push and a modicum of enlightened international collaboration to produce a cascade of energy-efficient items and techniques which could cut oil consumption in the advanced economies by a third in the next ten years and more soon thereafter. Some economies, such as the Swedish and the Japanese, are already on this path and the biggest consumer, the United States, is just waking to the fact that its proverbial ingenuity and private enterprise dynamism are well capable of meeting this challenge, given the right lead and the right public policy framework.

It is, incidentally, precisely because the Japanese are on track with end-use energy efficiency, and have stayed on track ever since the first oil shocks of the 1970s, that their industrialists have been prepared to take the plunge with a four or five year lead over the rest of the world, by investing and tooling up massively for new products such as the hybrid car, micro-grid technology and low-energy heat exchangers.

But it is by no means too late to follow the same path quickly and even to catch up somewhat. The biggest advances ready to be made are in the transport sector, which accounts for half the world's current oil demand and in the home heating and lighting sector. In the United States Professor Amory Lovins and his research team at the Rocky Mountain Institute have set out in immense detail the practical and early ways in which changes can be triggered and carried through in this area, transforming in the process both America's domestic industrial and social scene, and its present unhappy stance in world affairs.[88]

The central Lovins messages are:

1) that it is much cheaper **not** to use a barrel of oil than to use one;

2) the transition to much lower oil usage can be swift, smooth and profitable.

Reducing oil consumption by not using oil, and instead investing in alternative methods and already available technologies, makes big economic sense even if one takes only into account the market price of $50–70 plus. Add in all the colossal hidden costs of trying to safeguard and maintain oil supplies which fall on governments and taxpayers and divert national resources from better causes (including better energy use) and the saving becomes far, far greater.

The technologies now exist and are already being used which can deliver immense oil-use savings and do so quite rapidly. With state-of-the-art technologies, which lie just ahead, the savings could be even more dramatic, releasing enormous resources for domestic users who are now paying for imports and costs that disappear into petrodollars.

The Hybrid Car – the New Sputnik

Transportation is the sector, drinking up as it does 60 per cent of all oil consumed (higher in America), where the biggest changes are both needed and are attainable in a short timescale. Hybrid vehicles, both cars and trucks, which until a few years ago were just a very old technology gathering dust on the shelf, are now moving centre stage.[89] Toyota's Prius, which first appeared in Japanese markets in 1997, has been described as an 'industry sputnik' by one leading motor manufacturer[90] – a product and a harbinger of design revolution almost a decade ahead of its non-Japanese rivals. Only Honda has been able to keep pace. Belatedly the world's other automotive industries have begun catching up, with a big new range of hybrid vehicles being developed, including, most ambitiously, vehicles still with massive Sports Utility Vehicle (SUV) bodies, so beloved of American drivers, but with hybrid power trains.

The hybrid is spreading like wildfire in the USA, more slowly in Europe, despite the much higher tax on motorists' fuels, and slowest of all in the countries of the Middle East where gasoline and diesel are not only sold at cost but actually below cost (i.e. subsidized), so that the incentive to go for high-miles-per-gallon energy-efficient vehicles is zero. In central Teheran the cost of a litre of motor spirit is 14 cents (9 pence).

Hybrids are lighter than conventional cars. The gasoline engine in a hybrid can be much smaller than the one in a conventional car and therefore more efficient. Smaller engines are more efficient than bigger ones for several reasons. The big engine in a conventional vehicle is heavier than the small engine in a hybrid, as are other internal components, hence the car uses extra energy every time it accelerates or drives up a hill. Bigger engines usually have more cylinders, and each cylinder uses fuel every time the engine fires, even if the car isn't moving. Reducing the overall weight of a car is one easy way to increase the mileage.

A hybrid also recovers energy and stores it in the battery. In a conventional vehicle, braking is done by mechanical brakes, where the brakes remove some of the car's kinetic energy and dissipate it in the form of heat. Braking in a hybrid is controlled in part by the electric motor, which can recapture part of the kinetic energy of the car to partially recharge the batteries. It does this by using 'regenerative braking', that is, instead of just using the brakes to stop the car, the electric motor that drives the hybrid can also slow the car. In this mode, the electric motor acts as a generator and charges the batteries while the car is slowing down. This further reduces wear on brakes from the regenerative braking system used.

Hollywood stars by the score have purchased hybrids. High-profile hybrids drivers such as Leonardo Di Caprio, Billy Crystal, Harrison Ford, Kevin Bacon, George Clooney, Natalie Portman and Susan Sarandon have done a world of good for hybrid cars. In the UK various political leaders have bought Toyota Prius models in a haze of well-publicized virtue.

And why not? After all, the hybrid is a low-polluting and low-petroleum consuming car. It may cost more (even in America after the subsidy hand-out),[91] but it does seem cheaper to drive and in the London metropolitan area there is that wonderful feeling, not easy to cost, that the congestion charge no longer makes its daily intrusion – one more damn thing to remember![92]

An R.L. Polk survey of Year 2003 model cars showed that hybrid car registrations in the United States rose to 43,435 cars, a 25.8 per cent increase from 2002 numbers. California, the nation's most populous state at one eighth of the total population, had the most hybrid cars registered: 11,425. The high number may be partially due to the state's higher gasoline prices and stricter emissions rules, which hybrids generally have little trouble passing. Figures for 2005 and 2006 are far higher, with a three-year waiting list now in California for Toyota's latest models.

This unexpected increase in demand has surprised even Toyota, which has simply been unable to deliver in sufficient numbers. As a result, Toyota dramatically boosted its capacity to support the sale of 130,000 Prius in 2005, just over a tenth of them in Europe. Further production facilities are being prepared outside Japan, including in China. Mr Shinjiro Toyoda, the grand old man of the whole Toyota empire, told the authors that his firm planned to build a million hybrids a year by 2008 in Japan, and possibly as many again in Toyota's Shanghai plant.

How on earth did Toyota get so far ahead? The firm started its research in the hybrid field in 1965. Toyota Sports 800 launched in 1977 as a gas turbine hybrid prototype. Soon all Toyotas will come with hybrid engine options. The trend is clear: the supply of hybrid cars is set to rise very fast indeed. According to Kazuo Okatomoto, the man in charge of Toyota's research and development, design and product development, 'in 20 or 40 years all the automotive group's cars will be hybrids'.[93] It could be much sooner.

Several car manufacturers are now desperately scrambling to catch up, including General Motors, which initially had little faith in the hybrid solution. Ford and Nissan have entered into licensing agreements that allow them to use Toyota's hybrid technology.[94]

There are also plans to expand hybrid trains, trucks and buses. Toyota claims to have started with the Coaster Hybrid Bus in 1997 on the Japanese market. In 2003 GM introduced a diesel hybrid military (light) truck, equipped with a diesel electric and a fuel cell auxiliary power unit. Hybrid light trucks were introduced in 2004 by Mercedes (Hybrid Sprinter) and Micro-Vett SPA (Daily Bimodale).

Is the Hybrid Really a Solution?

While it is undeniable that hybrids use less fossil fuel to operate, hence emit less CO_2, the question that many people ask, not always in a friendly way, is how much extra fossil fuel the manufacturing of a hybrid and its components use?

One contention is that the net energy use by a hybrid, from its construction onwards, is actually higher than it would be for the non-hybrid version. This is due to the extra energy cost of the electric motors, wiring and, especially, the nickel metal hydride batteries which are not present in the non-hybrid car.

CNW Marketing Research released in 2006 a 450-page report on what they call 'Dust to Dust' automotive energy use. CNW tried to estimate all the energy used by an automobile over its life cycle – its initial design and development, its manufacture, through its lifetime of use, and its ultimate disposal. CNW suggest that hybrid vehicles may be part of the problem, not part of the answer: 'Put simply, over the "Dust to Dust" lifetime of the Honda Accord Hybrid, it will require about 50 per cent more energy than the non-hybrid version. One of the reasons hybrids cost more than non-hybrids is the manufacture, replacement and disposal of such items as batteries, electric motors (in addition to the conventional engine), lighter weight materials and complexity of the power package.' The main point seems to be that the energy used to fuel a vehicle is only a part (a very small part, CNW concludes) of the total energy required by that vehicle during its entire 'Dust to Dust' life cycle.

But in this kind of analysis of the 'full car life cycle energy use', the assumptions are undoubtedly questionable. For instance, CNW suggest that much of the energy used in a vehicle's life cycle is the energy needed for recycling and disposal. But even if true, that raises an accounting issue – should energy expended in recycling be counted as the end of the life cycle of the automobile, or as the beginning of the life cycle of whatever the recycled material is used for next?

The debate amongst experts will continue and rival technologies will come along. But the success of the Prius does indeed have sputnik qualities. It is a huge wake-up call and a reminder to forecasters and economic trend analysts that tech-nological innovation is always there, sometimes quietly waiting in a long-established concept, but ready to spring unending surprises and invalidate yesterday's assumptions and today's extrapolations. It was Margaret Thatcher who said (although she may not have been the first) that one should always expect the unexpected. Who had heard of hybrid vehicles 20 years ago? A handful of specialists and a few companies distributing milk trailers and golf buggies. Who has heard of them today? Just about everyone. The world can change itself radically in an amazingly short space of time.

The Bearable Lightness of Driving

Alongside hybridity the other big transport energy saver is lightness of vehicles (and for reasons explained above, the hybrid tends to be anyway a lighter car). Ultra-light cars and trucks could be manufactured now, but for the extraordinary conservatism of designers and the conviction – 180 degrees wrong – that lightness in vehicle construction means less safety.

Both these developments – hybrid power and ultra-lightweight construction, have their sceptics and their denigrators, and the technologies can certainly be carried much further. It is also the case that very modern diesels, designed to meet the latest US standards, can deliver mileage performances comparable with the current hybrids. But innovation is happen-ing by the hour and any measurements and tests are out-of-date almost before they are read. As of now hybrid power plus diesel refinement point an immediate way forward, which could make decisive inroads into oil consumption and oil imports.

Why does it not happen more quickly? Because the fact of high oil prices being here to stay has only just sunk in, because policymakers and market researchers are running ever further behind breathtakingly rapid events and because too many opinion-forming minds are still focused on the wrong priori-ties. Only a few more nudges from public policy would rapidly transform the scene.

Some are happening. On the American side the big nudge already took place, encouraging a hybrid as the new family car – namely a one-off $3,500 subsidy on every purchase from the Government. In the UK the annual vehicle licence duty is lower (£70 against £168 in 2006) and there is the relief in London from the burdensome congestion charge, soon to be raised to £25 a day for bigger and greedier, so-called 'off-road' vehicles. How this relief will be worked out for bigger 'off-road' vehicles which are nevertheless hybrids, already now marketed in America, will be interesting.

Planes and Trains

The gains in fuel consumption from using lighter materials, both advanced composites and lightweight steels, apply just as much to trains, ships and above all to aircraft. The world's passenger jet fleet is hugely inefficient compared with what is attainable and has already been attained in Boeing's newest aircraft like the 777 and the 7E7. These are around 20 per cent more fuel efficient than the average, and maybe 50 per cent more efficient (i.e. using less kerosene) than the queues of more aged craft which fill many of the world's fleets.

The irony here is that while airline operators cling to these ancient planes, some of them hand-me-downs from the big airlines which could be as much as 50 years old, their businesses are all being crucified by high fuel costs. As Warren Buffet remarked, airlines are a great industry but a bad business. Trapped by low profits, they are stuck with aged vehicles that drain away still more cash, while the capital to invest in far more efficient, but admittedly very expensive, new low-cost aircraft is just not there.

Public policy could ease this deadlock but has not yet done so. However, piling new taxes on air travel generally, to choke off customers, is probably the worst possible way to meet the problem. A war on cheap air travel is not the right way forward. For one thing, it is also highly regressive. As incomes rise above the barest minimum the freedom to travel by air, at low cost, is one of the great liberating forces of humanity.

Well-heeled policymakers have no right to take it out of reach, especially since it will have only marginal effect, and crowded in the margin will be the least well off. As taxes are piled on cheap air travel, with everyone in authority nodding in agreement, it just seems not to be understood in policymaking or political circles that in the modern world air travel, like car travel, is a basic freedom, a gateway to liberty from the constrictions of static life.

The European Union targets for cutting jet aircraft emissions sound fine. But it is the way they are implemented that matters. The best plan would be to bring in an innovative loan guarantee programme (as proposed by Lovins) to purchase or lease new fuel-efficient aircraft, or even to finance retrofits of new and more economical engines to older bodies. This would be linked to incentives to scrap old aircraft parked on remote runways and dragged back into service when business picks up. It would certainly be less expensive than waiting for airlines to go broke and bailing out their pension funds.

Another move which would help cut emissions far more effectively than piling taxes on passengers would be to accelerate investment in the most modern ground-handling facilities, so that aircraft would no longer have to circle aimlessly above hub airports, notably London's Heathrow, thus wasting fuel in large quantities and of course emitting carbon copiously.

As for train travel, the Japanese have set by far the best example here, using technologies which put Japan years ahead of most other industrialized countries. The aim, quite simply, has been to make domestic air travel between major Japanese cities redundant. This is why the internal Japanese airlines are already being squeezed by the high-speed, ultra-light, low-energy Shinkansen train sets, with their multiple drives (an engine beneath every set of bogies) and their hugely greater energy efficiency (and minimal carbon emissions).

Coming in the next few years is the proposed new Yamanashi Maglev (magnetic levitation) system, which has already been tested to the nth degree and will definitely go ahead, non-stop, between Tokyo and Osaka, although it could take two decades to reach full operation. It will be financed out of current operating profits – an indication of the massive success of the existing

155

high-speed network. At speeds of 450–500 kph its energy costs per passenger mile, and its carbon emissions per passenger mile, come in so far below the performance of jet airliners that there is really no comparison.[95,96]

The complexities and capital costs are high, although by burrowing deep beneath hills a great deal of planning delays and property rights squabbling is obviated. The London-Edinburgh route (about the same as Tokyo-Osaka) could be well suited to a similar project. Domestic flights to Scotland would probably cease, with very large savings in CO_2 emissions. Needless to say a Government-appointed committee, headed by a former British Airways Chief Executive, has turned the idea down. A Government seriously interested in energy security and longer-term carbon reduction would resurrect it as the UK's internal travel spine of the future. The train should replace not the car but the plane.

Oil and Armies

Oil is used massively by the armed forces as well. Information by fuel use by the army, navy and air force is hard to come by, but it is difficult not to believe that large energy savings could be achieved not only to ease stretched budgets but to revolutionize the warfare patterns of the past which relied so heavily on fuel supply lines. Inability to supply military transport played a large part in weakening German defences in Europe during the Second World War, as allied forces rolled forward, generously fuelled by petroleum shipped from the USA. The logistics required to get these fuel supplies forward (eventually General Patton's tanks were halted by fuel shortage) were, and are, themselves a formidable resource drag.

Quite aside from questions of agility in modern warfare, are tanks which consume a gallon of fuel every three miles (or three gallons every mile as some cynics claim) the right sort of equipment in an age when oil costs more than $50 a barrel? It seems extraordinary that the super-gas-guzzling Abrams tank, favoured by the US military, should be such a feature of American combat capability. The energy appetite of this 70 tonne monster is truly awesome. Each tank, swallowing up as little as three miles travelled per gallon, and gulping fuel at appalling levels of inefficiency even when its main engine is idle, has to be followed by numerous 5,000-gallon tanker trucks, themselves requiring in turn yet more fuel. These ponderous giants seem to symbolize the slow-moving inefficiency of American military might which becomes less and less able to project and sustain rapid power in a transformed and infinitely more agile world of weaponry and fighting operations. This is despite the proverbial bravery and toughness of the best American fighting units.

The Home Front

We could also be on the verge of other considerable energy saving methods in the home. The obvious immediate winners on this front are cool light bulbs, already available everywhere, and LED cold lighting, just coming into the shops and needing only a small push to become universally available at a competitive price in both installation and operating terms.[97] All the familiar advice about roof insulation is there for the taking, while new building standards in most European countries now insist on substantial energy-saving methods.

'Micro-generation' or 'embedded generation' are the buzz phrases here. In addition to the familiar items such as better insulation, double glazing etc., suppliers can now offer solar panels, now being manufactured with super-light-sensitive and efficient coatings, solar tiles, which are less obtrusive than large panels and look like ordinary roof tiles, or small wind power generators in the garden or on the roof. Heating from the ground and high-efficiency domestic equipment can all help.

Less discussed as yet in the UK is the revolutionary technology for replacing all domestic boilers, gas, oil-fired and electric, with heat pumps which cost about a quarter of conventional boilers to run, emit 32 per cent less carbon than a gas boiler and condense water to full washing and central heating levels. The initial capital cost is still deterringly high (the reason why most consumers talk a lot about energy saving but do not do much), but it could fall dramatically soon. The French already subsidize these types of installation and, given the cheapness (and low-carbon character) of French nuclear electricity, this leads to even more substantial energy savings and to shrinking instead of soaring utility bills in the home.

What Could be Done Now – Not in 40 Years' Time

The estimate is that overall oil use could be cut by half in the US in the next 18 years (by 2025), and imports by two-thirds. For the UK the timescale could be much shorter still. An easing of demand on this scale might not alter the oil price much for a while ahead – for reasons already explained to do with fast-rising production costs. But it would take the political heat right out of the system for the crucial bridging years – possibly in time to offset the impact of extreme oil price (and gas price) volatility.

Meanwhile, to achieve a reasonably quick and smooth transition to lower oil use, the moves, in which all the players, not just governments and not just the big oil-drinking nations, have a part, must begin in intricate sequence.

They commence with governments and with the shaping of coherent and clear supportive policies. But it must be emphasized that this is just the start and it is not governments who then take the lead. That role belongs to business enterprise which is already positioned to go ahead rapidly to make the transition faster, cheaper and more attractive in every market-place and region.

It can take this lead and set the pace if the public policy framework is supportive and thoroughly conducive to innovation and if subsidies, and if grotesque price distortions and protectionist devices are minimized in both energy supply and demand. Above all, to get the complex process of unwinding oil dependence truly under way, and to reduce the paralysing uncertainty facing all energy planners and investors, a revised international context is a necessity. The foreign polices of the leading powers must be re-thought, and international collaboration and cooperation between ALL parties, that is between

Tomorrow's pylons or yesterday's pylons – neither very friendly to the green and pleasant landscape.

consumer and consumer nations, as well as between consumer and producer nations, so desultory and inconclusive up to now, must be instigated.[98] In a speech in July 2006 in Chicago Tony Blair astonished opinion by asking for 'a renaissance of thinking' about foreign policy stances and aims, in relation to the Middle East in particular. It all seemed a bit late in the day, but perhaps the prodigal should be welcomed.

America will need to be first off the block because America is the biggest consumer by far, using up a quarter of the world's daily oil supply. Contrary to much mythology and thinly disguised anti-Americanism, which finds everything about the USA wrong and unpleasant, the US performance so far has not been at all bad. Between 1976, at the time of the first oil shocks, and 1985, as has been noted, the American economy delivered the staggering performance of a 27 per cent increase in GNP alongside a reduction in oil consumption by 17 per cent.[99]

For both good reasons and bad (i.e. efficiency and slowed world economic growth) the world market for oil shrank by one tenth over that period. That was not enough but it demonstrates that in a modest way the deed was done and could be done again, albeit on a much larger scale, and copied by the European and other already developed economies.

In a different form the pattern needs to be followed by the new big importer/consumers on the block, China and India. A vast rise in car ownership in China is inevitable. Not even a Communist dictatorship will be able to prevent the Chinese abandoning their bicycles for the freedom and liberation of their own motorized vehicles.

The European past performance is also an exemplar of what can now be done again on an even bigger scale. It is a surprising fact that Britain, France and Japan consume no more oil today than they did in 1973, despite having much larger GNPs. But it is still far, far too much. The pressures which prevented oil consumption going even higher therefore need to be understood, re-mobilized and now greatly reinforced.

The triggering steps by government – any government, but start with the UK – will be:

1) To think in terms *not* of replacing 'market failure' with more state intervention and additional tax burdens, but in terms of market correction to enable the powerful forces of enterprise and innovation to respond to new conditions and new needs.

2) To steer public choice, both on the part of businesses and private individuals (and public agencies), strongly towards advanced technology, very high mileage vehicles and trucks.

3) To practise what it preaches in the way of energy efficiency and conservation with its own large fleets of vehicles, but also throughout in the public sector generally, including the energy-thirsty military sector.

4) To encourage, or where relevant not to discourage, through planning laws and the lightest possible regulation the emergence of new industrial clusters producing the equipment and services demanded in a low-oil and energy-efficient age.

5) To back the right sort of innovative research which both boosts the best use of existing technologies and accelerates the emergence of new low-energy technologies.

6) To work internationally for the closest cooperation with all the other big oil-consuming countries. China and India should undoubtedly be members of the IEA, the Paris-based consumers 'club' which generates mutual support both in sharing stocks and other cushions in facing the next short-term oil shocks and in developing oil-saving techniques and technologies which can be practised worldwide.

7) To ensure that tax regimes and regulations do not get in the way of the most efficient and profitable development of energy resources within each government's domain. In the UK case that means setting North Sea oil and gas taxes at correct levels to ensure sustained high output and giving strong encouragement to the revival of the UK's indigenous coal industry and to new clean coal technologies.

8) To recapture national control of farm-support policies and use them to encourage the emergence of profitable, whole-chain, biofuel systems, from crop to pump. In Europe this means at last replacing the outdated Common Agricultural Policy structure which is conducive to massive energy inefficiency, high-cost farming and low profitability.

9) To avoid diverting overdue attention and debate away from the so-called 'solutions' to the energy issue which lie in the uncertain and distant future, and to give an authoritative and honest lead instead in handling the immediate challenges to the world energy balance which, if not addressed, will jeopardize both short- and long-term energy and environmental goals.

Existing European policies for maintaining high petrol (gasoline) taxes, imposing urban charges, imposing lower speed limits and more intensive traffic calming arrangements will all continue to help. But it is a delusion to think that high fuel costs via high fuel taxes will alone impel people to swap to higher mileage vehicles. A few more pence on a litre may marginally reduce journeys, so saving, say, £200 a year by getting a more efficient car may seem a nice idea. But since a swap typically involves borrowing and laying out another £5,000 to £10,000 or more, it takes a lot more than savings of this order to turn an idea into an act. Increasing taxes on the motorist from their already high levels in Europe (and especially in the UK) is also very regressive, as are taxes on air travel.

A war on car numbers on the roads, like a war on cheap air travel, is pointless, as well as heartless. Car ownership, and the freedom it brings to take the family on holiday, get the kids to school, chuck the pram in the back and carry home really heavy shopping, is a liberty so desired that it becomes almost a right which should not be denied. Public transport has its role, but only the well-off could believe that the poorer income groups could somehow be fobbed off with public transport.[100]

For government measures to bite, a good deal of willing compliance and recognition of the justice and commonsense of constricting rules has to be in place. That is why it is so vital that government leaders should speak frankly and tellingly about the true urgency of the oncoming crisis and the need to avoid the otherwise inevitable energy train crash.

Sweden has tried to develop a serious, government-led programme for basing all energy on renewable resources by 2020. How is this going to be achieved? The answer is that it won't be, or not at least by government measures, targets and hopes. The Swedish dream was, and still is, to build up renewable energy supplies, chiefly from the country's vast forests, and phase out not only all oil imports but nuclear power as well. But having announced a phase-out of nuclear power some years ago, and closed one big station, Sweden still finds itself relying on nuclear power for 34 per cent of its total energy (54 per cent of its electricity).

Renewables account for 28 per cent of Sweden's energy, but biofuels, on which the highest hopes in Sweden rest, have made little inroad into the transport sector and the Swedes still persist in driving round using good old gasoline. The Swedish energy 'model' is a healthy reminder that targets, official strategies and good intentions at government level are not enough.

Conclusion: The Power of Fashion (and Peer Pressure)

Given the right government steps on the energy efficiency front, and given good public information about what is possible and desirable (and economical), the near-term goal of a substantial reduction in world oil demand is well within reach. Governments set the framework, but markets, local initiative and enlightened self-interest will do the rest.

Ensuring that consumers pay the right and full price for fuel has to be the key driver in shifting the pattern of energy demand. But the next biggest motivation, and the one which seemed to have real effect in the energy saving periods of the 1970s and 1980s, is social pressure and social acceptance of the common danger both to energy supplies and to the environment, and the common and individual need for steps and actions to avert it. Grass-roots movements can be as powerful as, or more powerful than, market forces in bringing about lifestyle changes. There are, for example, definite signs that big, gas-guzzling vehicles like the notorious 'Chelsea Tractors' in London's smarter areas, are becoming socially unacceptable.[101] Latest sales for SUVs in the USA are reported to be 20 per cent down on the previous year.

Peer group pressure works around business boardroom tables as well as around kitchen tables. And industry in the advanced nations is the biggest consumer if not of oil then of gas and electricity, which creates indirect pressure on the oil scene. Companies, however, have bottom lines and backers and shareholders to satisfy. They cannot afford merely to be fashionable when they decide to invest other people's money. The biggest deterrent to investment in new energy-efficient technology and machinery is the fear that cheap oil will return, invalidating all cost-saving calculations and leaving white elephant projects stranded.

That is exactly what happened to many well-intentioned and enthusiastic businesses in the 1980s. The biggest incentive to move forward, to invest in not just the new methods but the new products which the public are already beginning to demand – in the home, in the kitchen, in the office, on the road, in the air and in all public places – is the certain knowledge that cheap and reliable oil will not come back for years, if ever, that the full price of both oil and other fossil fuels will have to be paid and that cheaper and much more profitable energy platforms are waiting in the wings to be rolled on stage.

CHAPTER EIGHT

ARIADNE'S THREAD: THE WAY OUT OF THE LABYRINTH

രൂ രൂ രൂ

The tentacles of geopolitics are wrapped around
energy supply. How the world can be made safer
for reliable energy. The central need for a re-shaped
international context. And how the policymakers
can, and must, act without delay.

രൂ രൂ രൂ

9th September 1980. Around the breakfast hour, in gaols
across the Kingdom of Saudi Arabia, 67 individuals were led
into the prison yards, forced to kneel and decapitated.

Sheikh Zaki Yamani, the Saudi Oil Minister, whispered this
news to the visiting UK Energy Minister, as their car sped that
morning from the hotel after breakfast towards the Palace for a
courtesy call on King Faisal.

This was the latest grim twist in the saga which had begun
with the capture of the Mecca Mosque the previous November
by Sunni extremists and a subsequent two-week siege which
French paratroopers had to be called in to end. These were the
67 leaders of the rebels who had dared sully the Holy Place and
had shaken the security of the entire Kingdom to the core.

The British Minister sat chilled and sickened for a moment. He could have asked what was the point of this terrible and brutal vengeance, but he decided to keep his thoughts to himself. For every head sliced off he suspected ten more would spring up. For every death there would be endless deaths in the endless future. This was a world in which violence and hatred, between religions, between sects, between power-holders and power-seekers, between the mega-rich and the third world poor, between the sleek villas and the squatters' shanties, between urban glitter and urban poverty, was endemic. It could only grow and grow. It contained two-thirds of the world's known crude oil reserves. But it was not a region on which his own people, or the West generally, could or should rely. There had one day to be a better solution to global energy needs. The problem was how to get there.

The Unlikely Future

The future is non-linear. That much we know. To expect things to carry on as before is invariably wrong. What the ancients called Fate always, but always, intervenes. Surprises, shocks, sharp bends in the track of events are a certainty, and sometimes a happy turn of events is a possibility. Fate can be kind as well as cruel.

It may be that everything will suddenly change for the better in the Middle East. It may be that Israel and Palestine will find a *modus vivendi*. It may be that Iran will start playing a positive role in preventing nuclear proliferation. It may be that the bloody civil war in Iraq will come to a halt. It may be that Shia will become brother to Sunni and each will live in harmony. It may be that democracy – in various guises – will spread sweetness and light through the region as kingdoms and emirates convert to parliamentary government. It may be that in Saudi Arabia opposition to the House of Ibn Saud will evaporate and the terrorists and extremists will fold their tents and depart. It may be that peace will descend on battered Lebanon, the Hezbollah fall quiet and retire into their villages and hills, and Beirut will rise again as the Paris of the East which it once was.

It may be, too, that a few years ahead a calm and orderly low-carbon energy world emerges, with plenty of renewable energy sources, with stable but declining fossil fuel supplies and with expanding and safe nuclear power feeding reliable and plentiful electricity into homes and offices and factories. All this is noble and possible and to be devoutly wished and worked for.

But we know perfectly well that it is all highly unlikely. Even for the short distance we can see ahead it cannot be. For those in authority to assume that any of these things will come about soon, and that the world can glide from here to there over the months and years immediately ahead, would be a total dereliction of duty and responsibility.

The Tentacles of Geopolitics

The outgoing British Prime Minister, Tony Blair, called in his final speech to his Party's annual conference for 'the most radical overhaul in energy policy in 50 years'.

But energy policy cannot be detached from its surroundings. There is no energy shortage, never has been and never will be. The risks, dangers and insecurities all arise in safely capturing, organizing, transmitting and delivering power to an expanding population at the points where it is needed and demanded, and in avoiding the tendrils and tentacles of geopolitics that wrap around the entire process. And the challenge is to stay free of all these within the timescale and within the environment in which human beings live and die, work and play, and love and learn.

Whatever happens, and however unwelcome it is to hear, the world will consume more and more fossil fuels for at least the next century. The responsible reaction is to plan for that and seek to avoid or soften the dangers. The irresponsible reaction is to deny it and dream of a different global order. Despite major increases in the efficiency with which energy is used, both current and achievable, world demand for energy will continue to grow massively as the billions in the not-yet-developed world not only attain the standards of the richer countries but in some case leapfrog them.

The main form in which power will be sought and delivered will be electricity. The future will be electric worldwide. That will be the main method of provision, whatever the primary source behind it. Where power is needed in mobile form the most convenient pattern will be hydrogen, which is hyper-plentiful everywhere around us, but needs big electric currents to separate it from oxygen in the air.

It is clear enough that this combination – of electricity universally available and hydrogen as the key to mobility – is within the reach of technologies now being developed and that it satisfies man's strong and understandable desire to offset the carbon emissions of past ages in favour of a better environment decades hence. The search for greater energy security and the search for a low-carbon future are interwoven, but the focus must be on both, with the former taking priority, otherwise both will fail.

Meanwhile, there is life to be lived and societies to be sustained and to be allowed to prosper, and this means not theorizing and dreaming about the future but facing the immediate and deadly problems which the energy scene presents. These demand action and international coherence of approach on an unprecedented scale if they are to be brought into harness and under control and if all the longer-term hopes for the welfare of the planet and its inhabitants are to be fulfilled.

This revision of the international context is supremely urgent. Wise American voices can be increasingly heard, insisting that in the United States, the biggest consumer of all, energy policy and foreign policy should be properly integrated and not treated separately, as they have been over past decades.

Now that the UK is again a major net energy importer it should be on the same path, bringing foreign policy and energy policy together, and in turn weaving longer-term climate goals, which can only be achieved by intense global cooperation, in with both.

That is why, although the energy debate and its accompanying literature are usually full of long-term strategies and visions, this book is unapologetically about the short-term, the here-and-now and the scene a little way ahead (and for most people even a year is a long time). Maynard Keynes was right. Not only are we all dead in the long term but if we are careless now, or badly misled, or captured by flawed theories and ideologies, there will be no long term worth handing on, or just a condition so grim and narrowed that future generations will curse their forbears.

So the message is to act now, to create an interdependent platform on which a good future can be built and in which calm decisions can be taken by individuals, by enterprises, by governments about new lifestyles and the least disruptive and painful ways towards them.

Action now means action to escape the oil trap and the frightening oil scramble it is leading to. In a sense the terrorist threat to world oil supplies has already done a power of good, by reminding the world that the situation just cannot go on.[102] Dependence, as at present for daily life, health and survival, and for the very existence of civilized society, on the narrow hatreds and vendettas of the world of Islamic extremism, is so risky, so distorting and so unnecessary that no governing authority which allows it to persist will be tolerated.

Oil is the centrepiece of the problems and transport is the centrepiece of the oil issue. If world transport consumption of oil can be halved over a decade that takes all the pressure out of the energy scene and allows for measured moves to a low-carbon energy future to gather momentum. Of course other developments on both the supply side and on the demand and use sides can help greatly. But basically it will all be marginal to the central oil question.

This is not a shortage issue. It is a matter of power, politics, religion, instability and terror. The Middle East will remain dark with violence and bloodshed for years to come. The huge power-dispersing effects of the information revolution, paving the way for e-enabled terrorism and group violence on a scale never seen before in history,[103] ensure that the tensions will increase, not diminish, into the indefinite future. World energy cannot be in thrall to these vagaries.

Power in New Hands

Practical, effective and realistic action has to take place at all levels – at the international level, at the national and governmental level, at the business level and at the grass roots consumer level of the individual. Perhaps that is putting it the wrong way round since it is individual people – in their homes or at work – who generate the demand by insisting on new methods, services and products, to which business enterprise and government then have to respond.

But even the most insistent demands from the marketplace and the grass roots of societies will be fruitless if international tensions and misguided foreign policies and postures are barring a constructive response. Today's unbalanced world and unending march of foreign policy errors guarantees the worst possible conditions for safe and secure energy supplies and for a smooth passage into a different energy paradigm. So repairing that has to come first.

At the core of international instability lies an unavoidable fact, although it is one which political establishments are reluctant to face. Power has been dispersed. The information revolution, beginning in the 1970s and still being propelled to ever more amazing advances in microchip technology and communication, has redistributed power in the world both between nations and between nation state governments and other entities.[104]

Not only is the centre of economic gravity gradually shifting from the Atlantic nations, America and Europe, to the rising Asian powers. The power, or capacity to dominate and impose certain courses of action on others, however unwilling, has passed in part from states and national authorities to groups and causes outside national control. E-enablement empowers new non-state groups in two sets of ways.

Small is Lethal

First, the mobile phone and the internet have given groups, whether large or very small, well-intentioned or malignly intentioned, immense scope, never before enjoyed, to plan and organize, mobilize and act with devastating impact and precision. Second, the miniaturization of lethal and ultra-effective weaponry has likewise put power to deploy violence straight into the hands of smaller groups. Armed power can no longer be measured by weight and mass, when tiny weapons can be programmed with deadly accuracy by a handful of operators to destroy with extensive impunity the ships, aircraft, tanks and other mega-machinery of war, as well as fixed installations however well protected.

In a sense the lesson is a very old one – that skilled guerrilla forces even with completely conventional weapons, maybe used unconventionally, can often outwit more cumbersome armies. In modern times America had a taste of this changed balance in the Vietnam War and has found the same again in Iraq. The British twentieth-century lessons were in Malaysia (where the lesson was well learnt), in Kenya, in Cyprus and in Northern Ireland. Most recently the Israeli military, armed to the teeth with the latest heavy weaponry which ought to have prevailed against all guerrilla groups, found to their dismay that their tanks were highly vulnerable and that the very weight of their advanced equipment was a handicap.

In short, information technology and weapons miniaturization have vastly reinforced the advantage of the small over the large, the Davids against the Goliaths. Small that was beautiful has become small that is lethal. Those states and institutions which held power under the old order now have to husband their dwindling stock of influence and use it in entirely new ways.

The failure of the Bush Administration in Washington to understand this new dispensation – either the world shift away from Atlantic dominance or the technology shift away from states – is near the root of the present energy dangers. The conviction that America by its size and immense military expenditure can always prevail, and that democracy can always be imposed by overwhelming force, has led the USA and its allies straight into the Middle-East quagmire.

After 1945 the towering statesmen of post-war America – Harriman, Marshall, Truman and Eisenhower – found themselves at the helm of the world's most powerful and prosperous nation by far. They used their position wisely and remembered the first lesson for the powerful in the democratic age – be humble and circumspect in public statements. There were moments, certainly, when the kid gloves came off, but for most of the time allies were solicitously sought, cultures respected, generosity, both to victors and vanquished, maximized and the obviously overwhelming position of America left discreetly untrumpeted.

America today, or its governors, has forgotten all that. Bush speeches bellow out US power and claimed strength, seemingly unaware of the nation's vulnerability. The Terrible Simplifier has been busy at work on his utterances, lining up the world between good and evil, between those who are for us or must be against us.[105] Assertions pour out from the Administration that America is the leader of the democracies and the free world, that it has the power to spread democracy (shallowly defined) to every corner of the globe and that it intends to use overwhelming force to do so, together with others if they care to come along but, if not, then alone.

It seems strange that a nation with such a broad and generous track record, and one which once commanded so much respect and admiration, should now be narrowing its vision so drastically and misunderstanding the new dynamics of the network world so utterly. Sometimes it seems as though a great chunk of American academia – and not just the neo-conservative wing – has been imprisoned in a flawed, or perhaps one should say outdated, analysis of the new international power realities, and this has in turn infected the Washington administrators. Having been educated to see a world of blocs and superpowers and hegemonies, and having heard voices round the world denounce American power, they now only believe what they hear and think they can see, which is an America which is all-powerful and should reform the planet.

It is not so much the obvious fact of American size and sheer military weight (the USA still accounts for over 25 per cent of the world's GNP and its defence budget is ten times that of the next nearest down the list – Japan), but the way it is handled and deployed. America is said to be in the grip of a religious revival, with church attendances soaring and the evangelists wielding considerable power over public opinion and, indeed, public policy. What America perhaps needs is a different kind of religious ethic, such as that embodied in

Japan's Shinto religion, whose central tenets are to be modest and to be moderate.

Yet without a moment's hesitation American officials step forward to 'take the lead' in resolving Middle-East issues, in handling prickly Iran, in pressurizing Syria, in bringing freedom to Iraq, in lecturing Russia (and others selectively) on democratic procedures and in instructing all other members of the United Nations on how to proceed in bringing order to an unruly world.

Thus the world hears Vice-President Dick Cheney laying into Putin's Russia for its harsh oppression and lack of democracy in terms of unparalleled diplomatic clumsiness, certain to aggravate but not to heal and without any understanding of the internal dynamics of Russian governance. The failure of American leaders to seize the opportunities offered by the end of the Soviet Union, and of the Cold War, and to build new and strong common ground with a fast reviving Russia, will be one more black mark going down in the history books against the second Bush administration.

Britain in Limbo

This is the kind of lead which now has no followers, except, perhaps, the outgoing British Prime Minister, Tony Blair. For Tony Blair, the dream of Britain as some kind of bridge between the United States and the European Union has crumbled and lies in a heap of concrete on the river bed. The design was attractive but the construction would never have held up. The anchor points either side were missing. On the American side, Middle-East policy has failed. On the European side, there is no unity of view on which to build.

On the US side, for all the protestations that Britain was, or is, America's trusted and equal partner, that was never the Bush team's view. It was nice to have Britain on side and Blair was a great guy, and so on. But the Americans needed no middle man to interpret Europe to them. They could see for themselves the virulent anti-Americanism in 'old Europe' and they anyway believed – wrongly, as it turns out – they could manage alone.

On the European side the rhetoric has been plentiful all along, but the reality remains slim. The EU's Common Foreign and Security Policy, which leading EU Ministers continue to describe daily as being essential to secure Europe's influence on the world stage and which the Foreign and Commonwealth Office in London still regards as the UK's foreign policy priority ('working through our European partners' etc.), is a feeble instrument and anyway little involved in protecting and promoting Britain's real interests, or enabling Britain to make its most effective contribution to global peace, stability and development, or to the vital matter of energy security.

The consequence of all this is that Britain's foreign policy remains in limbo, and Britain's capacity for influence in the Arab Middle East has been neutered, with central and critical implications for energy security and supply. A leader more experienced in international diplomacy than Tony Blair might have understood that between America's pious belligerency and the anti-American hostility of most of Europe, as well as Russia and China, lay a wise and constructive middle route, along which the UK was ideally qualified to lead. Tough diplomacy, as opposed to declarations of enmity to all who questioned the war on terror, might have avoided the string of disasters in the Middle East which have occurred.

But that was not Blair's way. Full and unqualified commitment to the American 'New Middle East' strategy was promised, with the backing of an equally inexperienced Opposition leader.[106] Just off stage, British oil industry chiefs could also be heard whispering encouragement to go along with the American approach and calling for full involvement in the Iraq venture; 'otherwise', so one captain of industry was reported, 'British oil companies would lose out to their American competitors if London did not participate in a war on Iraq'.[107]

Pax Americana Crumbles

Meanwhile, American influence, instead of being, as in the post-war world, quietly effective, is in its steepest decline in history. American presence and policy in the Middle East has become a prime cause of continuing turmoil. Governments which seem too ready to tolerate American intervention and pressure become themselves the targets for attack. Oil becomes identified as the liquid refresher of Western evil, as well, perversely, as the source of almost unlimited funding for the terrorist weapons arsenal. The oil supply chain therefore takes centre stage as the prime target for attack, ideally at its most sensitive upstream points such as Ab-Qaiq. As the Al-Qaeda high command sees the situation – to take one amongst many not-necessarily-linked extremist groups – oil is the lifeblood of the hated crusader cultures of the West. Attacking hub points in the oil supply chain therefore gives the fanatics the chance to inflict lethal damage.

An irony is that the bulk of Middle Eastern oil actually goes to Asian buyers and Asian markets, and not to the West at all. So a prime task for a more clear-thinking US Administration would be to cease 'assuming' leadership in the region, and share the responsibilities with the non-Western nations with most to lose from continued Middle Eastern chaos and most to gain from a stable world and an assured pattern of future energy supplies.

Who Fills the Vacuum?

This tragic decline of America's 'soft power', reputation and influence almost across the entire globe is leaving a dangerous vacuum. Into this vacuum, cautiously, subtly, but steadily, are moving not the Europeans, with their slow growth and their protectionist mentality, but the Chinese – with cash, with investment projects, with trade deals, secured access to oil and gas supplies in an energy-hungry world, with military and policing support and with technology. A replay in reverse of the fifteenth century is unfolding, when China retreated in on itself, forbidding its great 'red ships' to explore further, while Europe reached outwards to every corner of the planet. Now it is exactly the other way round.

The global vacuum is one which ought to be filled not by the Chinese dictatorship but by the free democracies of

175

Europe, Asia and Africa, from both North and South, banded together by a commitment to freedom under the rule of law and ready to make real and common sacrifices in the interests of a peaceful and stable world and the spread of democratic governance in many different forms.

One much underrated organization which could assist in meeting this new need is the Commonwealth (the former British Commonwealth). This amazing network, both govern-mental and non-governmental, of 54 nations, small, large and giant, like India, possesses the vital attributes for dealing with this new world which the old twentieth-century institutions so conspicuously lack. It stretches across the faiths, with half a billion Muslim members; it stretches across all the Continents, thus by its very existence nullifying the dark analysis of a coming clash of civilizations.

With its confidence (and resources) boosted by its members, which include 13 of the world's fastest growing economies, it could play a decisive part in creating the platform for stability and democracy the world now longs for, as well as offering a forum in which many of the energy problems of the day could be tackled cooperatively, and in ways helpful to the penurious developing countries.

Meanwhile, for wounded America the immediate need is for far closer and more 'equal-terms' links and dialogue with China, Russia, India and Japan, the other major oil consumers – taking the opposite line to the browbeating Mr Cheney and his Washington colleagues. These are the countries best placed, with the USA, to set the global agenda. In fact, they are already doing so. An American administration blinded by the belief, which it makes no effort to disguise, that the USA is indeed the hegemonic power and the natural leader of the world is going to find it difficult to face the fact that power has now slipped from its grasp and lies elsewhere, and that other nations and cultures must accordingly be treated with the utmost respect and spoken to on completely equal terms.

But American and other Western leaders should long ago have been pressing for the inclusion of China and India in the consumers' 'club', the IEA in Paris, and long ago recognizing the central need to build on common energy interests.

It is absurd, and a glaring failure of international cooperation policy, that China should be in a scramble for oil in rivalry
with other nations, when the common interest in secure and
long-lasting supplies for all is so obvious. Whatever else is
achieved by business, by the energy industries or by the individual home-owner and motorist, the first task of governments
is to defuse the oil scramble and bring international coherence
to energy resource issues.

Return of the Dialogue

The second international task for statesmen should be to rein-
vigorate the links between oil consumers and oil producers
since here, too, the common interests are obvious and strong.
Security of supply and security of demand go hand in hand. No
developed forum exists in which the interested parties can
gather and hammer out a common strategy. A possible candi-
date would be an expanded version of the International Energy
Forum, set up in 2003 to facilitate a Ministerial-level dialogue
between major energy producers and consumers and with its
headquarters in Riyadh.

A weak dialogue has indeed been carried on between the
EU and OPEC for some time,[108] but this is far too narrow on
both sides. The new forum should be neither Atlantic dominated (in other words by neither the EU nor the USA) nor
should it be dominated at the other extreme by anti-American
sentiment, which currently prevails at the UN. The USA should
be an important participant as it is fully entitled to be, both as
THE major consumer of oil and a major, although declining
producer. But it should be just one quiet and wise voice (as it
used to be), and not the noisy bully at the front of the class,
threatening everyone, including the teacher.

The third priority task for government leaders in all the
Western and the Asian capitals is to be truthful and frank
with their publics about the real nature of the immediate
energy threat. The temptation to focus on visions of the world
ahead, in which carbon emissions have been lowered to zero
and a better world unfolds, have proved irresistible to too

many leaders who should have known better. As Peter Tertzakian puts it: 'our politicians continue to perpetuate the belief that cheap fuel, clean environment, secure supply, discreet infrastructure and a competitive economy all go hand in hand'.

That's for tomorrow. Today the crisis is upon us. In the UK, for example, domestic gas prices rose 50 per cent in 2006. Gas companies blamed rising wholesale gas prices which they said had climbed by 87 per cent since the beginning of 2005. During that same period oil prices doubled.

There is much more to come and the policies and measures at government level, as part of a broader strategy involving all sectors of society, ought now to be being wheeled into place. For these events and trials, and the measures to address them, the public should long ago have been prepared.

The fourth priority for governments is transport policy. Since the oil price is the immediate driver and over-dependence on oil the main danger, it is extraordinary how little has been attempted, almost anywhere in the world, to make changes in public policy towards transport. Motorists are of course an immensely powerful lobby. All governments are wary of taking them on – and not just elected ones. The Chinese Government shows extreme caution in addressing the issue of gasoline prices and taxes, and this when Chinese car ownership is still in its infancy. As Chinese car ownership climbs to average world levels presumably the paralysis will be complete.

The saving grace here may be that the Chinese breakneck growth rate will not continue in linear fashion, as many forecasters suppose. The Government of the People's Republic of China is indeed in a fragile condition, with colossal disparities between a super-rich minority and a majority, in gigantic numbers, not only in the deepest deprivation but trapped in an ugly spiral of increasing environmental disaster, as rivers and water sources run dry (the Yellow River has almost disappeared, being replaced by salt water), and increasing social disasters as the full horrific implications of the 'one child only' come home and the 'get rid of baby daughters' consequence leaves villages womanless and families dying out. This is the potential nightmare of discontinuity which may yet 'solve' the problem of

Chinese energy demand by the most violent and miserable means.

Meanwhile, in almost every country which produces oil the political resistance not just to heavy taxes on gasoline but even to pricing gasoline economically is judged to be almost too strong to overcome.

Although higher taxes on gasoline will not check the universal irresistible impulse to possess a vehicle, proper economic pricing, reflecting both the market and hidden costs of getting fuel to the consumer, is essential. If policymakers feel too weak in individual countries to push through the necessary measures then let them try holding hands to give each other courage.[109]

All new cars and trucks being built and marketed should from now on be very high miles-per-gallon vehicles (which means hybrid or advanced-technology diesel). *All* vehicle tax regimes should favour high-mileage cars over gas-guzzlers and bangers. *All* manufacturers should be pressured into going over to lightweight materials (carbon and light steel) for vehicle building. *All* highways (motorways) should be furnished with intelligent highway systems to minimize unnecessary jams and delays. *All* oil companies should be required to mix a proportion of biofuel with their petrol (gasoline) sales at the pump. *All* vehicle engine makers and component makers should be required to make only products which are biofuel-enabled and mixed fuel-friendly (on the model of Brazil's 'flex cars').

The Leadership Question

Governments nowadays can and do contract out many tasks once thought to be the bureaucracy's monopoly. But what they cannot contract out is leadership. From presidents and prime ministers downwards, and especially from the chief executive himself or herself, must come 'a certain idea' of a nation's place, purpose, direction and contribution – and nowhere more so than in matters of energy supplies and the geopolitics with which they are intertwined.

Within the envelope of wise national leadership the men and women gathered in public agencies, around boardroom tables and around kitchen tables – that is both those who guide public bodies and companies and those who guide families – must take decisions about the future as best they can. It is a balance requiring the utmost skill and understanding and a balance which must give enterprise and vigorous innovation full scope to carry world society forwards. Every decision-maker knows that commitments to massive and long-term infrastructure systems and strategies can be made in an instant but fix the future for generations. Somehow these have to be matched with policy agility and the flexibility as technology roars ahead. How to commit without being trapped and locked in. That is the question.

Politicians have to take risks about what to tax, what to regulate, what to subsidize (the worst course but the easiest one). Business has to take risks about the amount to invest in tooling up for new products to present to the public and for which it hopes demand will emerge or be created, or a bit of both. Dozens of clean power innovations both for turning easily grown plants and other matter into fuel and for using energy twice as efficiently await the start signal. For individual consumers – heads of families, home-owners, motorists – the moment has come for rethinking the way they spend their cash and for screwing up courage to take some bold and different 'big item' decisions when they arise – e.g. on new cars, new heating systems, new bathrooms, new kitchens, new homes, new gadgets and comforts.

All this is ready to happen. But it will not happen unless those appointed to stand high take the initiative, call out the dangers, warn of the wrong turnings and act in clear-sighted and reassuring ways to help rather than hinder market forces.

Finale

A greener, cleaner and calmer future, with safe, reliable and affordable supplies of power and energy for each and every one is attainable. So, Yes, there can be optimism, and soundly-based optimism, not impractical idealism, at that.

But No, we are nowhere near that point. The path to better things is long and tortuous. Like John Bunyan's pilgrim, we have to pass through and survive many trials, tribulations and temptations, stretching over the years immediately ahead, to come to the happier state.

In short, there *is* a way out of the labyrinth, but Ariadne's thread winds and twists. There are many corners to be turned and difficult passages to be taken, some of which may seem at first sight to lead in the wrong direction, back into dark centre. And there are inviting-looking pathways which beckon towards escape but lead only to dead ends, or once again back to the starting point. Political vanities and policy errors lurk at every corner.

Even if we assume a calmer and less frightening international scene – and that is making a wildly optimistic assumption at the present time – there are bafflingly complex energy issues to be resolved. For example, it is evident that the route to an eventual and measurable reduction in greenhouse gases lies not through raising seductive hopes that the climate can be easily and swiftly changed (it cannot be), but through addressing with the utmost vigour the problems of energy security now confronting us.

Or take the call to eliminate all fossil fuel burning (that is oil, coal and gas). In practice, the route away from the precarious supply chain of volatile mineral oil may lie through turning first to other fossil fuels in new forms – cleaner coal and frozen gas. Cleaner fossil fuels may be the first and only step on the long march to a cleaner and more balanced planet, more secure from the ravages of violent climate change.

Or take the long-term hope for greener and more sustainable energy sources. The first part of the route to greater energy efficiency and a different energy mix may lie through fuels and technologies that are not so energy efficient, such as ethanol and not so spotlessly clean 'unconventional' oils. Tar sands from Canada, for instance, may be just the job for reducing dependence on oil from unstable Middle-East regions, but they are certainly not carbon-free. And paradoxically, the route to cheaper and reliable energy, a universal human need, almost certainly lies through higher energy prices.

The routes that are going to work are not the obvious ones, nor necessarily the most popular. Dire threats of global warming and its apocalyptic consequences will not be incentive enough to follow them. Utopian schemes for a worldwide regime of carbon pricing, administered by stratospheric agencies accountable to no one, merely distract. Much more compelling imperatives and much more direct incentives will be required. Indeed there are prior questions which just cannot be by-passed. Before taking the difficult and maybe painful escape route there has to be a reason for escaping at all.

With the world so full of energy sources, why the need to save energy? With the science of climate change so uncertain, why listen to the voices urging an exit from the labyrinth at all? Why should we believe the experts who have so often been proved spectacularly wrong in the past? Why take actions now to achieve outcomes half a century ahead? With oil a-plenty in the ground, why bother with the cost of changing lifestyles or habits, when there are so many more urgent problems confronting the world, such as poor medicine, bad education, water shortages and under-nourishment?

Today, just at this moment, the answers, or some of them, rest inside these pages and in this book. But at any hour they could rise from the print and spring fully armed into our homes, workplaces, communities, cities, societies and institutions. Will we be ready?

CHAPTER NINE

MAIN RECOMMENDATIONS AND MESSAGES OF THIS BOOK

1. The search for greater energy security, worldwide, and the search for a low-carbon future are interwoven. The focus must be on both, but with the former taking priority. Carbon pricing, rationing and offsetting schemes alone are too weak and uncertain to carry the burden.
2. International collaboration and coherence are demanded on an unprecedented scale, on both fronts – secure energy and efforts to slow global warming – if any longer-term hopes for a safer and more balanced world are to be realized.
3. Foreign policy and energy policy must be linked together by all responsible governments *now* and new independent platforms created on which nations can unite their common interests – such as the prevention of terrorist disruption of global energy supply networks.
4. The oil issue is at the centre of current world energy dangers, and oil consumption by all forms of transportation is at the centre of the oil issue. Practical, effective and prompt steps are available to check the further growth of world oil consumption radically.

Transport-related policies everywhere must be changed now, in richer and poorer countries alike. Market-driven alternative technologies and fuels are waiting in the wings, and in some parts of the world are already being used successfully.

5. Reduced oil dependence would transform all forms of power supply. Gas and biofuels can partly displace oil, and new alternative and sustainable sources, plus eventually safe nuclear power, can replace gas for electricity.

6. The power to act is not just in Western hands and not just in government hands. Power and influence have been dispersed mightily by the information revolution. Those nations which used to hold all the cards of power – the USA in particular – have now to play their hands in entirely new and much cleverer ways. World command and control through Pax Americana is no more.

7. The UK, by its compliance with US strategies, has placed its foreign policy in limbo and severely weakened its capacity to influence events – and to ensure energy security. It needs urgently to build ties with new friends – countries which are now setting the global agenda.

8. The dialogue between oil and gas consumers and the main oil and gas producer nations should be reinvigorated. They share the same interest in supply security and demand security. And the world's major consuming countries, old and new, viz. the USA, China, India, Europe, should collaborate much more closely in meeting both energy and climate and environmental challenges.

9. Courageous and candid political leaders and opinion-shapers should explain the immediate threats openly and frankly, and prepare their publics and electorates for big changes in energy habits, lifestyles and possibilities.

10. Around Cabinet tables, around boardroom tables and around kitchen tables, big decisions have to be taken now. The route forward is not obvious. Information is unreliable, opinions biased and guidance confusing. But there IS a way out of the labyrinth to a better, safer, greener future for our planet earth and for all its inhabitants.

NOTES

Chapter One

1 'Our Energy Challenge: Securing clean, affordable energy for the long term', Department of Trade and Industry, 23 January 2006, and 'The Energy Challenge' from the same Department, July 2006.

2 Keynes, John Maynard, *Tract on Monetary Reform*, page 64.

Chapter Two

3 Appointed Secretary of State for Energy in May 1979.

4 Between 1978 and 1987 the new CAFE regulations in the USA were destined to improve the fuel-efficiency of US-made cars and trucks, by far the biggest oil consumers, by two-fifths.

5 The Straits of Hormuz, at the mouth of the Arabian (Persian) Gulf, are 21 miles wide, but the navigable channel for tankers is only 3 miles. Mines left in this channel, or even the rumour of them, would paralyze shipping and cut off 18 per cent of the world crude oil supplies.

6 The 'old' seven sisters were:
Standard Oil of New Jersey (Esso). This later became Exxon, now ExxonMobil.
Royal Dutch Shell. Anglo-Dutch.
British Anglo-Persian Oil Company (APOC). This later became British Petroleum, then BP Amoco following a merger with Amoco (which in turn was formerly Standard Oil of Indiana). It is now known solely by the initials BP.

Standard Oil of New York (Socony). This later became Mobil, which merged with Exxon to form ExxonMobil.

Texaco. This later merged with Chevron and was ChevronTexaco from 2001 till 2005 when the name of the company reverted back to Chevron.

Standard Oil of California (Socal). This became Chevron.

Gulf Oil. Most of this became part of Chevron, with smaller parts becoming part of BP, and Cumberland Farms. A network of stations in the north-eastern United States still bears this name.

As of 2005, the surviving companies are ExxonMobil, Chevron, Shell, Total and BP.

The rest have been swallowed up or squeezed out. Still more 'consolidation' could be on the way (see Chapter Two).

7   Car ownership in China is currently rising at about 15 per cent per year. Some 30 million cars are now on China's roads.

8   The Energy Information Agency within (although independent from) the US Department of Energy estimates a Chinese import figure of 10 million barrels a day by 2012.

9   See Peter Tertzakian's ground breaking book *A Thousand Barrels a Second*, McGraw Hill 2006.

10   See article 'The Way Ahead' by Dr. Carole Nakhle, published by The Society of Petroleum Engineers, February 2005.

11   The process by which gas is cooled and compressed to a liquid, shipped in tankers to consumer country destinations, and then warmed, re-gasified into its original form and pumped into local grids.

12   The Kyoto target for all signatories to the Protocol is to cut greenhouse gas emissions by 12.5 per cent on 1990 levels throughout the period 2008–12. The UK self-declared target is to cut emissions by 60 per cent by 2050. They are currently rising.

13   Between 2000 and 2004, US emissions growth fell by 8 per cent. EU emissions growth increased by 2.3 per cent. Source: UN Framework Convention on Climate Change.

14   US oil imports rose up to 1977, then dipped to about 25 per cent in the early1980s, then rose again steadily to the present level of more than 60 per cent.

15   Approximately, so Saudi officials say, 262 billion barrels.

16   Lovins, Amory B. *et al.*, *Winning the Oil Endgame* (Snowmass, Rocky Mountain Institute, 2004).

17   Angola, Azerbaijan, Colombia, Kazakhstan, Mexico, Russia, Nigeria, Venezuela.

18   *Winning the Oil Endgame*, ibid.

19   Page 201, *The World is Flat*. Penguin, Allen Lane, 2005.

20   See 'IEA and OPEC Must Come Together'. Article by David Howell and Carole Nakhle in the *International Herald Tribune*, August 2005.

21 George Soros of course teaches us that all markets, and financial ones in particular, overshoot, and that equilibrium is not a natural state. But for the reasons given earlier, oil may be out in front.

22 See, for example, Daniel Yergin's opinion: 'The years of past oil crises have demonstrated that, given time, markets will adjust and allocate'. *The Prize*, page 775. For a time they did, but eventually they didn't.

Chapter Three

23 For instance, Sir John Mogg, the UKI Energy Industry Regulator, told the *Financial Times* on August 24 2006 that gas supplies would improve in the winter of 2006. He even hinted that prices might fall, but was canny enough to add a warning about 'unforeseen events'.

24 In the UK it comes out at 70 per cent of the final price.

25 At the May 2005 International Monetary Fund gathering in Washington the UK's Chancellor, Gordon Brown, told bemused OPEC Ministers 'When OPEC meets on June 1st it must look at its production quotas and how we can increase both output and refining capacity'. No one had explained to him that the taps were already full on and that it takes four years minimum to build new refineries matched to the oil qualities available.

26 See *A Brief History of the End of the World*, Simon Pearson, Robinson; *The Last Generation*, Fred Pearce, Eden Project Books; *The North Pole was Here*, Andrew Revkin, Kingfisher Publications; and dozens more – all the time!

27 Said to have occurred in about 1450 BC when the nearby volcanic island of Santorini blew up in one of the largest explosions in world history. Not only was Minoan civilization devastated, over the whole of Egypt the sun was blotted out for at least a week and ash and debris (including frogs) were rained down on Upper Egypt. Vast tsunamis also sucked the tides out and threw back gigantic destructive waves. Moses, fleeing from Egypt at just about this time, may have been one of the lucky beneficiaries of this phenomenon.

28 Much respected – and very vocal – chair of the UK Government's Sustainable Development Committee.

29 2050 is a favourite 'demanding target' date. What is the betting that when the world gets there, if it does, the planners will have a new 'long term' set of targets to aim for?

30 As they have clearly done in California, where Governor Schwarzenegger has introduced legal limits capping carbon emissions. The consequent heavy penalties on high energy users may help his grandchildren. But will they help with the next oil shock?

31 *The End of Oil,* by Paul Roberts; *Twilight in the Desert,* by Matthew Simmons; *The Party's Over,* by Richard Heinberg; *Beyond Oil: the View from Hubbert's Peak,* by Kenneth Deffeyes; to name a few.

32 Sweden's official plan is to phase out oil usage by 2020 and replace most energy sources, including car fuel, with wood, of which it has a huge and renewable indigenous supplies.

33 The EIA forecast in 2003 a world consumption level in 2006 of 80 million barrels a day – at least four million too low.

34 Amidst struggles over oil wealth and security, of which the Darfur horrors in Western Sudan, as noted in Chapter One, are an ugly part.

35 Petrol in Teheran is 9p a litre. In Kuwait it is free!

36 See Hernando de Soto, *The Mystery of Capital. Why Capitalism Works in the West and Fails Everywhere Else,* 2000. de Soto has pioneered thinking about the key role of property rights in capital accumulation and development.

37 EU member states agreed in 1997 to aim by 2020 for a global average temperature of no more than 2°C above pre-industrial levels, as estimated at the time. That means a concentration in the atmosphere of less than 550 parts of CO_2 per million. There are doubts whether even that would do the trick. EU Heads of Government agreed in March 2005 to 'explore' targets of a 30–50 per cent CO_2 reduction by 2020, and 60–80 per cent by 2050.

38 IEA, 2005 World Energy Outlook.

39 Neither the USA nor China nor India have signed the Kyoto protocol, setting carbon reduction targets.

40 See also the penetrating report from the House of Lords Select Committee on Economic Affairs 'The Economics of Climate Change', July 2005. Committed carbon crusaders could scarcely bear the telling realism of this report, which urged that more study should be made of the economics of climate change and pointed out that global warming could have positive as well as negative effects.

41 For example, some leading asset managers have put their names to a statement claiming that 'investment decisions taken now will have a big impact on current and future greenhouse gas emissions and the world's climate'. (Reported in the *Financial Times,* 6 October 2006.) This is peddling a false prospectus. There is no way that investors can affect the *current* world climate. They will have to wait at least 30 years to do that. The current climate has been determined by people and businesses decades ago when they had no idea what they were doing.

42 The Carbon Trust was set up as a state-funded agency in 1999 to 'promote carbon reduction in the UK'. Since 2002 UK carbon emissions have been rising. The chairman nevertheless paid himself a bonus of £45,000 in 2005–6. See Carbon Trust Annual Report 2005–6.

43 The Christian 'crusader kingdom' of Jerusalem lasted from 1050 to 1215, before being overrun by Saladin and the last knight crusaders ejected from the region.

44 *Winning the Oil Endgame; Innovation for Profits, Jobs and Security*, Amory B. Lovins and others, Rocky Mountain Institute, 2005.

45 Though recent new finds off Louisiana may check this.

46 With apologies to Jim Hacker, Anthony Jay's immortal *Yes Minister* creation.

47 Our nervous friend might also ask why the UK has been left with such a disastrously small gas storage capacity by past planners. Were they all caught by surprise that the UK's North Sea gas had been used up so fast, and that a return to heavy reliance on imported gas lay just ahead? Apparently so.

48 The Shanghai Cooperation Organization (SCO) is an intergovernmental organization which was founded on 14 June 2001 by leaders of the People's Republic of China, Russia, Kazakhstan, Kyrgyzstan, Tajikistan and Uzbekistan. Except for Uzbekistan, the other countries had been members of the Shanghai Five; after the inclusion of Uzbekistan in 2001, the members renamed the organization.

Chapter Four

49 See Chapter Two, page 21.

50 The Ghawar oil field, discovered in 1948, is the largest in the world. 280 kilometres long and 16 km wide, the column of oil within it is 300 metres high. Despite continuous production since 1951, it still contains 70 billion barrels of oil, twice the total reserves of the USA!

51 Figures from the US Department of Energy.

52 Al Moneef (2006). And 1200 billion in 2006 (see above).

53 Footnote Chapter One, ibid.

54 Ministry of Petroleum & Energy (2006).

55 When Russian Foreign Minister Sergei Lavrov was asked by one of the authors in November 2005 whether Russia planned to use its energy resources as a diplomatic weapon, he replied 'Of course'.

56 By Dallas-based energy reserve auditors DeGolyer & MacNaughton, whose clients include leading Russian energy companies such as Gazprom and the now nationalized Yukos.

57 US sources continue, with classic insensitivity, to refer to it as the Persian Gulf.

58 Quoted by Daniel Yergin in the April 2006 edition of Foreign Affairs.

59 Shell's decision to change the methodology by which reserves were booked, involving the eventual need to reduce stated reserves by 25 per cent, led to a huge internal row and the departure of several senior company figures.

60   Professor Peter Odell. Lecture to IAEE World Annual Conference, Potsdam, June 2006.
61   These are ExxonMobil, BP, Shell, and Total out of the original Seven.
62   Economist Oil Review, Section 2005.
63   Putting the capacity of the new field at 27 mm cf (764,554 cm) per day, pus 932 barrels (a day) of condensates. The new field's capacity could go up to above 50 mm cf (1.4 mm cm) per day.

Chapter Five

64   viz *The World is not Enough*, with Piers Brosnan and Sophie Marceau, 2003.
65   The so-called Commonwealth of Independent States – a sort of ex-Soviet Union old boys club of countries, with just a hint of coercion.
66   See, for example, the article in the *International Herald Tribune*, 6 May 2006, by David Howell and Carole Nakhle, 'Time for Dialogue with the Producers'.
67   The world oil industry more or less began in Baku in the late nineteenth century and vast oil fortunes were built up there – amongst others by the Rothschilds and the Swedish Nobel dynasty.
68   Abkhazia and South Ossetia.
69   Quoted by Lutz Kleveman in his seminal study of pipeline politics in Central Asia, *The New Great Game*, Atlantic Books, 2003.
70   President Bush after 9/11.
71   Quoted again by Lutz Kleveman.
72   A document drawn up hopefully by Western officials seeking Russian agreement at the 2006 St. Petersburg Summit, and urging full gas liberalization, including open access to Gazprom's pipelines.

Chapter Six

73   Global nuclear capacity is 400 gigawatts out of a grand total of 3,980.
74   UNDP Report, 2000.
75   Smil, 2003.
76   Fulton *et al.*, 2004.
77   One gigajoule equals approximately the amount of energy consumed by a 100 watt bulb over four months.
78   *Beyond Oil: The View from Hubbert's Peak*, by Kenneth Deffeyes, Hill and Wang, 2005.
79   'The EU Strategy on Biofuels: from field to fuel', November 2006.
80   Manufactured both by the Germans in the Second World War, through the Fischer-Tropf process, and later by the South Africans at Sasol plants.

81 Notably in the House of Lords at Westminster by Lord Ezra, a former chairman of the National Coal Board, and Lord Jenkin, a
former UK Energy Minister.

82 July 2006.

83 There are three routes at the moment – using enriched U235, generating plutonium in a nuclear reactor (both used respectively in
the Hiroshima and Nagasaki bombs) or using nuclear fusion to
make a hydrogen bomb – the H-bomb – tested but not yet dropped
on anybody.

84 Such as the so-called 'pebble' technology, invented, like so much
else, in the UK many years ago but recently taken up by the
Chinese, who are pressing ahead with a large programme of
nuclear station building.

Chapter Seven

85 *Winning the Oil Endgame*, ibid.

86 Japan and Sweden have sent even sharper messages to oil producers
who try to overprice. Between 1970 and 2007 Sweden has cut its
oil dependence from 70 per cent to 30 per cent. Over the same
stretch Japan has reduced oil dependence from 80 per cent to 40
per cent.

87 How the internet revolution was going to change almost all
aspects of human activity and life patterns, including energy
supply and its broader political context, was set out in 1999 in
David Howell's book *The Edge of Now – New Questions for
Democracy in the Network Age* (Macmillan). Political and
economic columnists and commentators who did not read it were
still ooh-ing and aah-ing eight years later at the ubiquity of the
internet's impact.

88 *Winning The Oil Endgame. Innovation for Profits, Jobs and
Security*, by Amory B. Lovins and colleagues, Rocky Mountain
Institute.

89 Hybrid-electric cars were invented by Ferdinand Porsche in 1900.
They overcome the mismatch between engine power and tractive
load by turning the wheels by a mixture of power transmitted
either from the gasoline engine or from an electric motor.

90 The Sputnik was the world's first rocket launched space satellite,
put up by the Soviet Union in 1958. It jerked the West into the realization that they were falling behind in missile and space
technologies and gave new impetus to American space technology.

91 Every purchaser of a hybrid vehicle in the US received, until
recently, a $3,500 payment – a so-called feebate – from the Federal
Government. This manna from the federal heaven has now been
phased out although tax reliefs remain.

92   In the Congestion Charge area of Central London, now enlarged, hybrid vehicles are exempted – currently – from the £8-a-day levy. When this goes up to £25 for larger-engine vehicles the advantage for hybrids will be even greater – unless it is removed.

93   Automotive hybrid technology became successful in the 1990s when the Honda Insight and Toyota Prius became available. The Honda Insight is designed to get the best possible mileage. It is a small, lightweight two-seater with a tiny, high-efficiency gas engine. The Insight weighs less than 1,900 pounds (862 kg), which is 500 pounds (227 kg) less than the lightest Honda Civic. It has the best EPA mileage ratings of any hybrid car on the market. The Toyota Prius is designed to reduce emissions in urban areas. Its gasoline engine only starts once the vehicle has passed a certain speed. The Prius has been in high demand since its introduction. Toyota has sold a cumulative 318,500 hybrids between 1997 and 2004, including 278,700 Prius, and Honda has sold a total of 81,867 hybrids between 1999 and November 2004. In 2005 more than 600,000 hybrid vehicles were sold worldwide. By 2006 this figure had doubled again.

94   Newer designs have more conventional appearance and are less expensive, often appearing and performing identically to their non-hybrid counterparts while delivering 50 per cent better fuel efficiency. The Honda Civic Hybrid appears identical to the non-hybrid version, for instance, but delivers about 50 US MPG (4.7L/100Km). For 2007, Lexus is offering a hybrid version of their GS sport sedan dubbed the GS450h, with a power delivery 'well in excess of 300hp'. The 2007 Camry Hybrid has been announced and is to be launched in late spring as a 2007 model. Also, Nissan announced the release of the Altima hybrid around 2007.

95   Independent figures from the Japanese Ministry of Land, Infrastructure and Transport state that the Maglev system will use half as much energy per passenger as air travel and emit less than a third of the CO_2 per passenger. If access time is included it will also be one hour quicker than air travel between Tokyo and Osaka – 1 hour 30 minutes against 2 hours 34 minutes. The existing Tokaido Shinkansen system (bullet trains) emits one tenth of the CO_2 of aircraft per passenger carried.

96   The Transrapid accident in Germany with a Maglev technology train in September 2006 was said to be entirely due to human error. If so, it does not invalidate the claims for the Maglev as a superior and super-energy-efficient transport method.

97   A light-emitting diode is a semiconductor device that emits incoherent narrow-spectrum light when electrically biassed in the forward direction. Taipei now has all-LED traffic lights.

98   Both the G7 Venice Summiteers in Venice and the G8 Summit in St.
     Petersburg 26 years later called in their respective communiqués for
     a closer dialogue between consuming nations and the OPEC pro-
     ducers. It happened in a desultory way but without genuine
     cooperation resulting.
99   In 1985–86. See Chapter One.
100 Official figures put car ownership in the UK in 2006 at 29.6 million
     – just under one vehicle for every two people. In effect, everyone of
     driving age now wants, and feels the need for, a car.
101 In New York the situation has not been helped by zealous legisla-
     tion requiring bulky child safety seats in all vehicles, so that if
     parents want to do a school run for,say, four kids they have no
     choice but to buy a bulky, and probably gas-guzzling, vehicle – a
     glorious example of the law of unintended consequences.

Chapter Eight

102 See article by Howell and Nakhle in the *International Herald
     Tribune*, 9 February 2006, 'Thanks, Iran, for the Reminder'.
103 As forecast in *The Edge of Now,* by David Howell, published in
     2000.
104 *The Edge of Now*, ibid.
105 The phrase 'the Terrible Simplifier' comes from the pen of the his-
     torian and philosopher Jacob Burckhardt, who saw him at work
     throughout nineteenth-century Europe. See *The Letters of Jacob
     Burckhardt.*
106 Iain Duncan Smith.
107 As quoted in the *Guardian*, 30 October 2002.
108 And every G7 (or 8) Summit communiqué for years past has called
     for closer dialogue between oil consumers and producers.
109 China, Saudi Arabia and Venezuela all regulate and subsidize their
     consumption of gasoline and other oil products. Russia charges its
     domestic gas consumers $42 per 1000 cubic metres, but its European
     export customers up to $300. How long can this continue?

BIBLIOGRAPHY

Blanchard, Roger D., *The Future of Global Oil Production: Facts, Figures, Trends and Projections by Region* (Jefferson NC, McFarland, 2006)

Clark, William R., *Petrodollar Warfare: Oil, Iraq and the Future of the Dollar* (Gabriola Island, BC, New Society Publishers, 2001)

Crichton, Michael, *State of Fear* (New York, HarperCollins, 2005)

Deffeyes, Kenneth, *Beyond Oil: The View from Hubbert's Peak* (New York, Hill and Wang, 2006)

Economist, The, 'Oil Survey', London, April 2005

Fried, Edward R., and Schultze, Charles L., *Higher Oil Prices and The World Economy* (Washington DC, Brookings Institution, 1975)

Gore, Al, *An Inconvenient Truth: The Planetary Emergency of Global Warming and What We Can Do About It* (USA, 2006).

Grant, Lindsey, *The Collapsing Bubble: Growth and Fossil Energy* (Santa Ana, Seven Locks Press, 2005)

Heinberg, Richard, *The Party's Over – Oil, War and the Fate of Industrial Societies* (London, Clairview Books, 2003)

Heinberg, Richard, *The Oil Depletion Protocol* (Gabriola Island, BC, New Society Publishers, 2006)

Howell, D., and Nakhle, C., 'OPEC & the IEA have to get it right this time', *International Herald Tribune*, 30 September 2005

Howell, D., and Nakhle, C., 'Thanks, Iran, for the reminder', *International Herald Tribune*, 9 February 2006

Howell, David, *The Edge of Now. New Questions for Democracy in the Network Age* (London, Macmillan, 2000)

International Energy Agency, 'World Energy Outlook' (2006)

Jaccard, Mark, *Sustainable Fossil Fuels – An Unusual Suspect,* (Cambridge, 2005)

Kleveman, Lutz, *The Great New Game – Blood and Oil in Central Asia* (London, Atlantic Books, 2003)

Kynge, James, *The Rise of a Hungry Nation* (London, Weidenfeld & Nicholson, 2006)

Leggett, Jeremy, *Half Gone – Oil, Gas, Hot Air and the Global Energy Crisis* (London, Portobello, 2005)

Lovins, Amory B. *et al.*, *Winning the Oil Endgame* (Snowmass, Rocky Mountain Institute, 2004)

Nakhle, C., 'Black gold: still a man's world?' *Oil, Gas & Energy Law* (2005), Vol. 3, Issue 3

Nakhle, C., 'Look at the North Sea in a new way', *Financial Times,* 28 November 2005

Nakhle, C., 'Managing decline: getting the North Sea fiscal regime right' (2005), *Oil, Gas & Energy Law* (2005), Vol. 3, Issue 2

Nakhle, C., 'Taxing a declining province', *Petroleum Review* (2005), Vol. 59, n.704

Nakhle, C., 'Liberalisation and energy security: competing or complementary objectives?', IBL Focus No. 30 (2006)

Nakhle, C., and Howell, D., 'Fuelling a climate breakthrough', *International Herald Tribune*, 15 November 2006

Nakhle, C., 'Is the oil industry still attracting the talent it needs? The way ahead' (2006), Society of Petroleum Engineer (SPE), February, Vol. 2, n.1

Nakhle, C., *Petroleum Taxation* (London, Routledge, 2007)

Norwegian Ministry of Petroleum and Energy, http://odin.dep.no/oed/english/bn.html

Odell, Peter, *Oil and World Power* 8ed (Harmondsworth, Penguin, 1986)

Phillips, Kevin, *American Theocracy: the Perils and Politics of Radical Religion, Oil and Borrowed Money in the Twenty-first Century* (New York, Viking, 2006)

Roberts, Paul, *The End of Oil* (London, Bloomsbury, 2004)

Rutledge, I., *Addicted to Oil* (London, I.B. Tauris, 2004)

Sampson, A., *The Seven Sisters* (London, Coronet Books, 1976)

Simmons, Matthew, *Twilight in the Desert* (Hoboken NJ, John Wiley, 2005)

Soros, George, *The Age of Fallibility* (New York, Perseus, 2006)

Stern, Jonathan, 'The new security environment for European gas: worsening geopolitics and increasing global competition for LNG' (Oxford Institute for Energy Studies, 2006)

Stern, Jonathan, *The Russian-Ukrainian Gas Crisis of January 2006* (Oxford Institute for Energy Studies, 2006)

Stern, N., *Stern Review on the Economics of Climate Change*, London (October 2006)

Tertzakian, Peter, *A Thousand Barrels a Second* (New York, McGraw-Hill, 2006)

Yergin, Daniel, and Stanislaw, Joseph, *The Commanding Heights* (New York, Simon and Schuster, 1998)

Yergin, Daniel, *The Prize* (New York, Simon and Schuster, 1991)

INDEX

Note: page numbers with (n) attached refer to notes.